D1669548

STEREOSPECIFIC POLYMERIZATION OF ISOPRENE

Acad. dr. ing.
ELENA CEAUŞESCU

POLIMERIZAREA STEREOSPECIFICĂ A IZOPRENULUI

EDITURA ACADEMIEI
REPUBLICII SOCIALISTE ROMÂNIA
Bucureşti 1979

STEREOSPECIFIC POLYMERIZATION OF ISOPRENE

by

ELENA CEAUŞESCU

PERGAMON PRESS

OXFORD · NEW YORK · TORONTO · SYDNEY · PARIS · FRANKFURT

U.K.	Pergamon Press Ltd., Headington Hill Hall, Oxford OX3 0BW, England
U.S.A.	Pergamon Press Inc., Maxwell House, Fairview Park, Elmsford, New York 10523, U.S.A.
CANADA	Pergamon Press Canada Ltd., Suite 104, 150 Consumers Rd., Willowdale, Ontario M2J 1P9, Canada
AUSTRALIA	Pergamon Press (Aust.) Pty. Ltd., P.O. Box 544, Potts Point, N.S.W. 2011, Australia
FRANCE	Pergamon Press SARL, 24 rue des Ecoles, 75240 Paris, Cedex 05, France
FEDERAL REPUBLIC OF GERMANY	Pergamon Press GmbH, Hammerweg 6, D-6242 Kronberg-Taunus, Federal Republic of Germany

First English edition 1983.

British Library Cataloguing in Publication Data

Ceauşescu, Elena.
Stereospecific polymerization of isoprene.

Translation of: Polimerizarea stereospecifică a izoprenului.
Includes bibliographies.
1. Isoprene. 2. Polymers and polymerization.
I. Title.
QD305.H5C4213 1982 547'.41 82-18096
ISBN 0-08-029987-3

Translated from *Polimerizarea stereospecifică a izoprenului* published by Editura Academiei Republicii Socialiste România, Bucharest, 1979

Filmset by Speedlith Photo Litho Ltd., Manchester
Printed in Great Britain by Express Litho Service (Oxford)

Foreword

I am very glad to be able to welcome the English edition of the book by Dr. Elena Ceaușescu, Member of the Romanian Academy, Fellow of the Royal Institute of Chemistry, on the very important subject of the *Stereospecific Polymerization of Isoprene.*

My interest in scientific progress in Romania is recent, grown out of a visit that my husband and I made three years ago. This was formally organized around my participation in an international conference, but it led us to explore a very beautiful country with many remarkable ancient buildings and to experience the warm hospitality of the people around us.

At the end of our stay I was able to meet Dr. Elena Ceaușescu, to see the great Institute of Chemistry in which she works and to have with her and her colleagues extremely interesting discussions on the scientific problems before Romania, their achievements in the past and plans for the future. The book now published provides an account of the particular part Dr. Elena Ceaușescu played in the solution of one important problem, the production of synthetic rubber for use in Romania. Dr. Elena Ceaușescu is now a First Vice Prime-Minister of the Romanian Government, President of the National Council for Science and Technology of Romania, Chairwoman of the Scientific Council of the Central Institute of Chemistry and, not the least important, President of the Romanian National Committee "Scientists and Peace". She is also the wife of the dynamic President of Romania, Nicolae Ceaușescu.

It was impressive to learn that while duly coping with these tasks she still manages to do research work in her laboratory.

She has written two recent books of a general character, *Research Work on Synthesis and Characterization of Macromolecular Compounds* (1974) and *Recent Research in the Field of Macromolecular Compounds* (1981). The present volume, first published in Romania in 1979, is more specialized, concerned with the outstanding scientific preoccupations of the author herself. It has been published already, or is in the process of being published, in Italy, Greece, Switzerland, China and the USSR. An English edition will cement the friendship we formed with Dr. Ceaușescu when she visited this country and will be very useful to all those interested in polymers — and in Romania.

The discoveries of Ziegler and Natta on the use of certain catalysts, particularly complexes of titanium and aluminium, to effect stereospecific polymerizations revolutionized the synthesis of long-chain polymers and were celebrated by the award to them of the Nobel Prize. Elena Ceaușescu herself began work in the field for a doctorate in chemical engineering. In the first part of the book, essentially her thesis, she prepared a survey of the relevant literature and then set out to investigate the conditions for the stereospecific polymerization of isoprene that would affect the yield of *cis*-1,4-polyisoprene as near as possible identical with natural rubber. The thesis ends with the production in a pilot plant of enough rubber to make tyres which were mounted on cars and tested by driving on various road surfaces. They lasted well. The second part is concerned with the large variety of experiments and tests necessary to derive all the details of an original process for the manufacture of synthetic *cis*-1,4-polyisoprene rubber on an industrial scale. It ends before the construction of the successful factory subsequently built.

I am not equipped myself with enough technical knowledge of the field of this work to give a critical scientific evaluation of its contents, but even a necessarily brief reading makes one think that the field of research surveyed by the author is vast and recent. There are 635 scientific communications listed from all over the world, East and West. Science here appears one and integrated, scientists in rapid and easy communication with one another. Even many of the technical processes used in the field soon become known to all concerned. One would of course like international exchange of such details to be even more complete — something which may be promoted in the peaceful world for which Dr. Elena Ceaușescu also works. And Dr. Ceaușescu's career itself is very remarkable. She sets out to study one of the most fascinating scientific problems of our time, was able to guide its use for practical purposes, to organize, at increasingly higher governmental levels, manufacturing processes necessary to her country. Briefly in the last section of the book she mentions problems for the future and particular ecological problems — how to deal, for example, with polymer wastes which accumulate year by year. There is no shortage of these problems for all the young, very intelligent scientists growing up under her guidance, if all the needs of the people of Romania are to be met and the country preserved still as beautiful as we saw it.

I offer sincere best wishes to my distinguished colleague, to all scientists there, along with my warm congratulations to Dr. Elena Ceaușescu for her outstanding achievements.

DOROTHY CROWFOOT HODGKIN

Preface

THIS book is a study of the synthesis of *cis*-1,4-polyisoprene rubber, a most important elastomer among the numerous types of synthetic rubbers, and one whose structure and properties are similar to those of natural rubber; *cis*-1,4-polyisoprene is widely used in the manufacture of tyres and general rubber goods.

Given the complexity of the subject, the area of research is vast and many studies have been made to elucidate certain theoretical aspects related to the reaction mechanism and kinetics, and the influence of various variables on the course of the polymerization reaction as well as on the properties of the polymer. The Ph.D. Thesis, presented in Part I of this book, focused on these particular problems.

The data gathered by the author during an extensive experimental research programme are presented in a systematic form in Part II of this book. These data were intended to serve as a basis for a new and original manufacturing process of *cis*-1,4-polyisoprene rubber on an industrial scale. The first large output unit was subsequently built and was characterized by high efficiency and improved technical and economic performances.

Contents

PART I

PART II

PART I

DISSERTATION FOR THE DEGREE OF DOCTOR OF CHEMICAL ENGINEERING

Bucharest, 1967

LITERATURE DATA

Introduction

THE chemistry and technology of macromolecules with elastomeric properties have been largely developed during the last 20 years. The ever higher demand for rubber goods on the world market, as well as the wide range of properties required for various uses, has stimulated a marked increase in the quality and variety of the synthetic rubbers produced. A brief survey of the history of this development would necessarily include a great number of aspects, such as the type of monomer used, the catalyst or initiator systems, the kinetics of the polymerization reaction, the different structural features of the resulting polymers, their behaviour in chemical and mechanical processing, as well as other technological and economic problems. In the introductory section of this work only some of the more important problems concerning the manufacture of butadiene rubbers, butadiene–styrene rubbers, butyl rubbers and polyisoprene will be discussed.

Earlier polymerization techniques using alkali metals were abandoned in favour of emulsion polymerization methods. Polymerizations with various Ziegler–Natta-type catalysts [1–4], organolithium compounds, lithium metal [5–10], cobalt compounds and alkylaluminium halides [11] have been discovered and studied since 1954, resulting in stereoregular polymers. At the same time, progress has been made in the structural analysis and characterization of macromolecular chains by physical and chemical methods, especially when related to the nature and the arrangement of the monomer units, their spatial configuration (both geometrical and optical), the average chain length, the possibility of free chain rotation, the number of branches, etc. The results obtained in stereospecific polymerization in recent years facilitated the manufacture, on an industrial scale, of a number of hydrocarbon polymers with elastomeric properties and a stereoregular structure. The use of stereospecific catalysts for the polymerization of isoprene makes it possible to control the reaction to obtain a stereoregular polymer, having a structure and characteristics similar to natural rubber. The results obtained in the synthesis of *cis*-1,4-polyisoprene permitted the design and construction of large industrial units.

Among the stereoregular rubbers, *cis*-1,4-polyisoprene is the most important since it closely duplicates the structure of natural rubber and can replace it in most uses because the mechanical properties of the vulcanizates resemble those of natural rubber. Research work on the manufacturing of *cis*-1,4-polyisoprene rubber is still proceeding on a world scale in order to get more information on certain theoretical questions, since the reaction mechanism and kinetics are not thoroughly understood even now. Another major interest in such studies is of an economic nature. Research work is focused on new simplified methods of monomer synthesis to reduce the cost of *cis*-1,4-polyisoprene rubber. Other investigations are concerned with new catalyst systems able to improve both the polymer characteristics and the polymerization process.

Since *cis*-1,4-polyisoprene rubber is so important as a substitute for natural rubber, this work studies the synthesis of *cis*-1,4-polyisoprene from isoprene of over 99 % purity as well as isoprene of *ca.* 25 % purity by using heterogeneous catalyst systems of the $AlR_3 + TiCl_4$ type, where R is ethyl or isobutyl. Attempts have been made to find correlations between the characteristics of the catalyst system, the reaction conditions and the properties of the resulting polymer.

The reagent purity is highly important in stereospecific polymerization since impurities may partially or completely inhibit the catalyst activity, changing the polymerization mechanism and kinetics. Therefore, investigations on reagent purification were performed in order to find the highest permissible limit of impurities which would not interfere with the polymerization process.

Among the stereospecific catalyst systems used in diene polymerization, the heterogeneous catalyst system showing a higher activity and providing higher stereoregularity and rate of polymerization was chosen. Two heterogeneous catalyst systems have been used, namely triethyl aluminium and triisobutyl aluminium complexed with titanium tetrachloride, which differ only in the alkyl group attached to the reducing metal.

An important role in stereospecific polymerization is also played by the type and concentration of both the monomer and the solvent, affecting the polymerization mechanism and the reaction kinetics. Therefore, isoprene of over 99 % concentration, obtained by the dehydrogenation of isopentane, has generally been used in polymerization experiments. To reach such a high concentration, successive purification steps are needed, involving increased costs. Polymerizations were performed in a solvent (a saturated aliphatic hydrocarbon). Both the monomer and the solvent were purified before polymerization in order to remove all impurities interfering with the polymerization process.

We have also investigated the possibility of using an isoprene fraction with only about 25 % monomer (rather than an isoprene of over 99 % purity) obtained via a simplified dehydrogenation of isopentane. No expensive purification operations are necessary to prepare this monomer. Only those components which interfere with the stereospecific polymerization process need be removed. The monomer contains isopentane, which is used as a solvent in the polymerization reaction. The polymerization of the 25 % isoprene fraction was carried out with the same $AlR_3 + TiCl_4$ catalyst systems used for the polymerization of isoprene of over 99% purity. The same correlations between the catalyst systems, the reaction conditions and the polymer properties were again studied. The polymerization of the 25% isoprenic fraction has not been reported in the literature and was therefore patented. Apart from a theoretical interest, this method offers certain economic advantages, considering the simplified synthesis of the monomer and the polymerization conditions provided by the catalyst system used.

The two *cis*-1,4-polyisoprenes obtained from the isoprene of over 99% purity and that of *ca.* 25% purity were analysed by physical-chemical and physical-mechanical methods in order to determine fully the characteristics of the resulting polymers, to establish their elastomeric properties and the possibility of using them as substitutes for natural rubber.

Chapter I

General comments on stereospecific polymerization

1. Research prior to the discovery of stereospecific polymerization

STEREOSPECIFIC polymerization is one of the greatest conquests of modern chemistry. The synthesis of such polymers is based on Ziegler's discovery [12] of the polymerization of ethylene at low pressure with catalysts containing a metal alkyl (aluminium alkyl) and a transition metal halide, namely titanium tetrachloride. The preparation of synthetic macromolecules with a regular chemical and steric structure was a difficult problem; once the first difficulties had been overcome and the stereospecific catalyst complexes had been discovered, the preparation of stereoregular polymers and even of new classes of polymers became possible.

Although intensive research work in this field did not start before 1953, there were previous attempts and a theoretical approach had been suggested.

In 1932, Staudinger [13] obtained crystalline poly(propylene oxide) and assigned the properties of that polymer to the regular structure of the chains. Apart from the first synthesis of a crystalline polymer, Schildknecht and his co-workers [14–17] published in a series of papers their results on the ionic polymerization of vinyl ethers with boron trifluoride etherate. They assigned the particular mechanical properties of these polymers to their crystallinity and postulated a steric order for the structural units within the macromolecular backbone. The steric order suggested by Schildknecht corresponds now to the isotactic and syndiotactic configurations of poly(vinyl isobutyl ether).

The syndiotactic structure of this polymer was determined by X-ray analysis and it was proved that the main chain takes a helical configuration, each coil containing three structural units.

Cooper [18] reported, even before Ziegler's discovery, the high stereoregularity of some vinyl polymers, such as poly(vinyl chloride), poly(vinyl alcohol) and poly(acrylonitrile). These polymers had relatively short chains. Their stereoregularity was attributed to the high polarity of the substituents causing the formation of stereoregular polymers even when conventional catalysts were used.

Apart from these few incidental syntheses of macromolecular compounds with a regular steric and chemical structure, there are several other papers dealing with macromolecular stereoregularity [19–23].

Apart from these papers dealing with the possibility of preparing stereospecific polymers, and giving hypothetical structures, investigations were made on the use of organometallic compounds or of some of their complexes for the polymerization of olefins.

Kraus [24] indicated that mixtures of inorganic halides and alkyl-aluminium halides were efficient catalysts for olefin polymerization. Hall and Nash [25] isolated alkyl halides and aluminium alkyls from the reaction of aluminium chloride and ethylene. As early as 1943, Fischer [26] was able to achieve ethylene polymerization at low pressure by using a mixture of aluminium, aluminium chloride and $TiCl_4$. Based on subsequent research in this field, Coover [27] showed that what actually happened in Fischer's experiments was the alkylation of aluminium to organoaluminium compounds, which further formed an active catalyst for the polymerization of ethylene by reaction with titanium tetrachloride.

The most important contribution to the synthesis of organometallic compounds was, however, made by Ziegler and his co-workers, who worked for many years on the synthesis of such compounds [28]. Starting from Ziegler's discovery in 1923 of a method for the preparation of organic compounds of alkali metals [29], a series of syntheses and research investigations were carried out, culminating in the discovery in 1953 of the polymerization of ethylene at low pressure. Studies dating as far back as 1927 are known on the polymerization of butadiene with lithium alkyl [30–36] or with lithium metal [37], then of ethylene with butyl lithium [38] or with triethyl aluminium [39–41]. Polymers with relatively high molecular weights could be prepared by the so-called "step-wise organometallic synthesis".

The discovery of the synthesis of organolithium compounds [42] via the reaction:

$$RCl + 2Li \rightarrow RLi + LiCl$$

permitted the preparation of higher, linear, lithium alkyls from the reaction between ethyl lithium and ethylene. The chain growth is, however, restricted in this reaction. The addition of ethylene involves high temperatures at which the lithium alkyl decomposes and is deactivated.

From the above experimental data, it has been concluded that lithium hydride and lithium alkyls are catalysts for ethylene polymerization, yielding higher α-olefins. This assumption was experimentally confirmed by Ziegler [39] who succeeded in preparing dimers with a well-defined structure from α-olefins, using organometallic compounds. Linear ethylene oligomers were prepared in a homogeneous system in the presence of metal alkyls. Such reactions could not be performed by the use of cationic-type catalysts.

Another important discovery (made in 1949) was the reaction of aluminium hydride with ethylene, another stepwise synthesis yielding aluminium alkyls capable of reacting further with ethylene to give higher aluminium alkyls [35, 39, 40–45].

$$AlH_3 + 3CH_2 = CH_2 \xrightarrow{60\text{–}80^\circ C} Al(C_2H_5)_3 \xrightarrow[100\text{–}120^\circ C]{CH_2 = CH_2}$$

$$Al\begin{cases} CH_2(CH_2-CH_2)_xCH_3 \\ CH_2(CH_2-CH_2)_yCH_3 \\ CH_2(CH_2-CH_2)_zCH_3 \end{cases}$$

In this way, higher α-olefins could be prepared from ethylene. If, instead of ethylene, AlR_3 reacts with α-olefins (propylene, 1-butene), dimerization is the main reaction [39, 41, 46–49].

When the reaction of ethylene with $Al(C_2H_5)_3$ was carried out in the presence of traces of nickel or by using nickel acetylacetonate, 1-butene was isolated instead of higher aluminium alkyls [25].

Ziegler later discovered that other catalysts (cobalt, platinum) show the same property [12, 46, 50–54].

At temperatures lower than 100°C, the reaction is much more rapid if a transition metal derivative is also present in the reaction mixture.

Research [55, 56] which followed the discovery of the so-called "nickel effect" started a systematic investigation to find other metals showing the same effect in the above reaction. By using zirconium acetylacetonate as the second component, Ziegler obtained polyethylene [12]. He later on discovered that certain titanium compounds, namely $TiCl_4$, are the most efficient.

Natta [57] suggested, by comparison with cationic polymerization, that the transition metal derivative be termed the catalyst, and the metal alkyl, the cocatalyst. His work played an important role in developing the polymerization reaction with Ziegler-type catalysts. He succeeded in polymerizing propylene and other α-olefins [58] and his first results on the crystallinity of these polymers were published in 1958 [59–62].

Natta and his co-workers [58, 59, 61–64] studied the preparation of polymers with a highly regular structure. By changing the catalyst system, they succeeded in obtaining with these modified Ziegler-type catalysts highly stereoregular poly-α-olefins, without using successive extraction procedures.

2. *Structural characteristics*

Stereospecific polymerization processes lead to head-to-tail linear polymers. The structural units of such polymers, with equivalent steric configurations, are oriented in a highly ordered sequence [64]. The reaction conditions of stereospecific polymerization largely favour a certain reaction pathway. The term "stereospecific polymerization" applies to the formation of a "tactic" polymer from a mixture of stereoisomeric monomers.

In their earlier works, Staudinger [13] and De Boer [65] assumed that the lack of crystallization of many polymers was related to the irregular spatial distribution of substituents along the main chain. After the discovery of stereospecific polymerization it became possible to synthesize polymers with a high degree of crystallinity. An ordered sequence of the structural unit configurations is determined by the presence of a chain backbone with the same configuration or an ordered sequence of enantiomorphous configurations. The "tactic" polymers may be partly or even completely amorphous, but always show a high crystallizability. It seemed at that time that synthetic polymers with a 100% degree of crystallinity could not be prepared [57, 66].

Stereoisomerism in polymers is highly important from a practical point of view since the most interesting and useful physical and technological properties are related to their crystallizability.

The chain conformation in the crystalline state, the crystallization rate, the size, distribution and orientation of crystallites are highly important in this respect [64].

Two types of isomerism can be involved in stereoregular polymers:

- geometrical isomerism, i.e. different geometrical configurations corresponding to the same chemical formula, for instance in compounds having double bonds or in cyclic compounds;
- optical stereoisomerism, i.e. a molecule can take enantiomorphous configurations despite free rotations around the single bonds; this type of isomerism is usually related to the presence of asymmetrical carbon atoms [67].

The steric order of the main chain was called "tacticity" by Natta and Huggins [68]. Therefore, the classification of polymers according to their steric order is obvious. The different arrangement of asymmetric carbon atoms, dextrorotatory and levorotatory configurations, produces structural modification of the macromolecular backbone. Natta [59, 60] was able to separate and synthesize polymers with a high degree of stereoregularity and, hence, crystallinity, and called the various configurations of the "tactic" polymers "isotactic" and "syndiotactic", respectively.

3. The main types of stereoregular polymers

Vinyl monomers, with the general formula $RCH{=}CH_2$, form polymers wherein the R substituent may be sterically placed in different arrangements [57, 63, 69–72].

The isotactic configuration defines a structure with all structural units taking a *d* or *l* configuration and all substituents placed on the same side of the main polymer chain, assuming for the latter a planar zig-zag configuration.

The syndiotactic configuration defines a structure with the structural units taking an alternating configuration of *d* and *l* optical isomers, the substituents being placed alternatively above and below the main polymer chain.

Figure 1 shows the three possible structural isomers of poly-α-olefins, namely isotactic, syndiotactic and atactic.

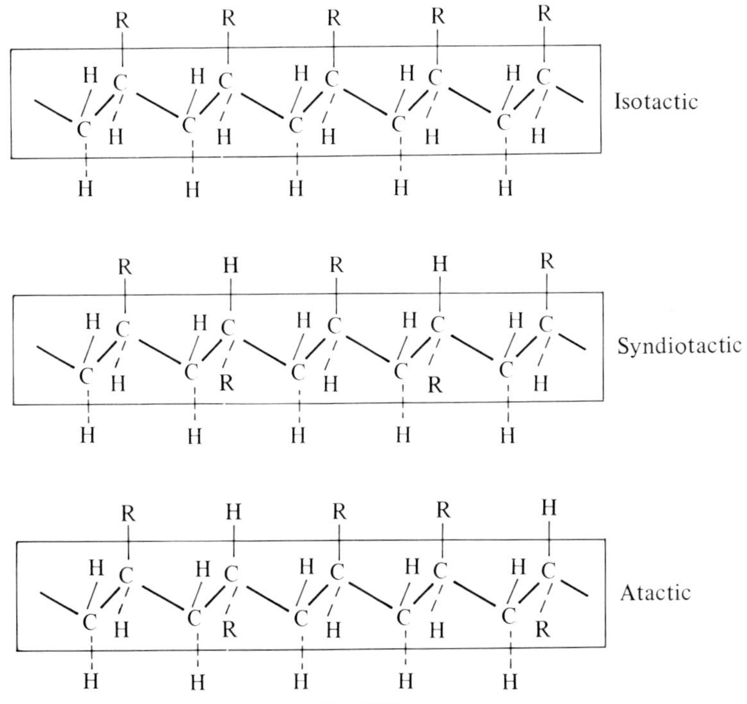

FIGURE 1. Structural forms of poly-α-olefins [73].

The type of stereoisomerism is dependent on the positioning and double bond rupture of each incoming monomeric unit during its addition to the growing chain [73, 75].

The 1,2-disubstituted ethylenes, $R{-}CH{=}CH{-}R'$, however, show a special type of tacticity arising from the presence of the two substituents.

The isotactic polymer may occur in two forms depending on the relative position of R and R′ (*cis* or *trans*). Hence, two types of polymers called "diisotactic" with different structures may result, designated by the terms "*threo*-" and "*erythro*-diisotactic". Such structures were made evident in the case of 1-methyl-2-deuteroethylene [74].

The polymer structure is given in Figure 2.

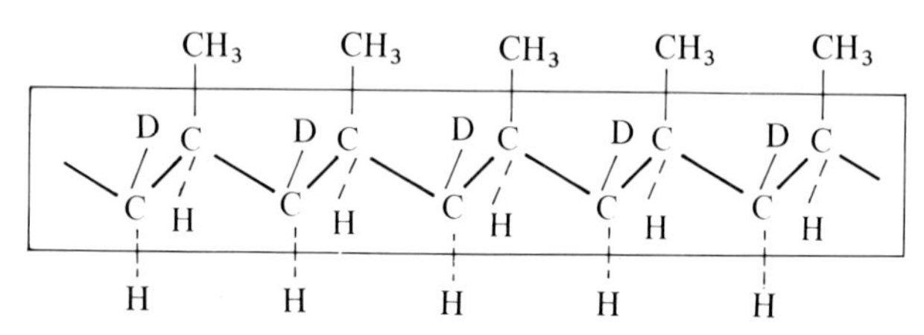

Threo-1-deutero-polypropylene
(or diisotactic *threo*-1-deutero-poly-propylene)

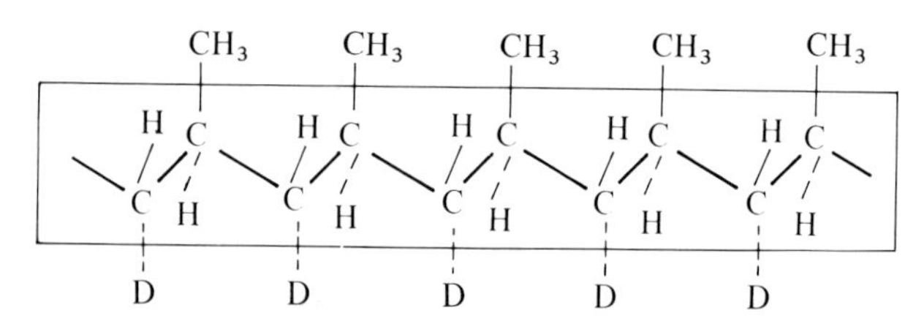

Erythro-1-deutero polypropylene
(or diisotactic *erythro*-1-deutero-polypropylene)

FIGURE 2. Di-isotactic polymers [73].

The polymers have the same helical structure as isotactic polypropylene, but the relative orientation of the D and CH_3 groups leads to this new type of isomerism.

A *threo*-diisotactic polymer is a diisotactic polymer in which both carbon atoms in the structural unit have the same configuration. The two substituents, R and R′, namely CH_3 and D in the case of 1-deuteropropylene, lie on the same side of the plane of the main polymer chain. Conversely, if the two substituents are on opposite sides of that plane, the polymer is *erythro*-diisotactic.

4. General methods for the synthesis of stereoregular polymers

After the discovery of stereospecific polymerization, intensive research work began with the aim of undertaking the polymerization of most monomers by using this process.

The question of stereospecificity and the influence of the catalyst type arises especially in the case of the polymerization of substituted olefins and dienes. According to Natta [57], in order to prepare sterically pure polymers, the following conditions have to be met:

— polymerization must occur in such a way that the monomer substituents be attached only in the 1,3-position (head-to-tail);
— there must be no branching of the backbone chain;
— the catalyst must incorporate the basic structural unit into the growing chain.

One condition for a monomer to undergo stereospecific polymerization is that during the initiation reaction and then again during the propagation stage, it reaches a given state, further maintained during the whole propagation reaction stage, which will favour a certain isomerism of the backbone chain. This orientation can be achieved in several ways [67]:

— by monomer adsorption on the catalyst surface, immobilized in the steric configuration in which it will be incorporated into the growing chain. Small molecules, such as ethylene and propylene, require a very strong complexing ability to form stereoregular polymers. Such strong complex formation can be achieved only by an active site adsorbed on a solid surface [76].

Butler assumes [77] that certain polar substituents, such as hydroxyl [28, 29], chloro, nitrile, and fluoro groups, favour the formation of polymers having long parallel chains capable of forming a network-type structure. The small size of the substituents allows an ordered arrangement of the polymer macromolecule with respect to a plane of pseudosymmetry:

— with homogeneous catalysts, the monomer forms complexes, involving the solvent or not as the case may be.

Higher olefins can be sterically controlled even by counter-ions in a homogeneous system, using the appropriate solvent and temperature to yield a complex.

Ethylene, substituted with bulky polar groups, may be converted into stereoregular polymers even by free radical polymerization provided the appropriate solvent and temperature are chosen, so as to favour a particular transition state [72, 76, 78–80].

In order to undergo a stereospecific polymerization, it is obvious that the monomers play a part in directing the polymerization mechanism, by stabilizing a carbanion, a free radical or a carbonium ion. With catalysts able to stabilize all these three types of active centres, the mechanism is controlled by the monomer and the polymerization conditions [81]. The monomer or the structural unit has to assume a certain isomerism (positional, structural or geometrical) corresponding to a given arrangement [60, 82, 83].

When Ziegler–Natta-type catalyst systems are being used, the monomer must contain electron-repelling groups complying with the tendency of the central atom to fulfil its coordination number.

Ethylene, and all olefins in general, have a relatively accessible and easily 'activated' electron pair, the π double bond electrons, which are involved in the formation of such types of complexes [84]. As mentioned before, strongly polar monomers, such as vinyl monomers with bulky substituents or with a given configuration, demand a minimum of conditions to undergo stereospecific polymerization.

The first diene-type stereoregular synthetic polymers were synthesized in 1954. *cis*-1,4-Polyisoprene was obtained with Ziegler–Natta-type catalysts [1–3] at the Goodrich Laboratories. The same rubber, "Coral", was manufactured by Firestone by using lithium metal [5–7], whereas Natta [85, 86] obtained *trans*-1,4-polyisoprene with a Ziegler-type catalyst.

Apart from stereoisomerism provided by the asymmetrical carbon atoms, conjugated dienes may also display geometrical isomerism, as the result of the *cis*- or *trans*-configuration of the double bonds present in the structural units.

1,3-Butadiene yields a polymer with a 1,2-structure when the polymerization occurs on the vinyl bond. Isotactic or syndiotactic structures are formed depending upon the substituent position (either on the same side of the backbone chain or regularly alternating on both sides of the backbone chain) [87].

The 1,4-addition-type polymers are formed when both double bonds are broken, resulting in the formation of a structural unit of the $-CH_2-CH=CH-CH_2-$ type. Each structural unit still contains a double bond with a *cis*- or a *trans*-configuration [88] (Figure 3).

Compared with butadiene, isoprene has one more possibility for isomerization during polymerization, because of its CH_3 substituent, i.e. 3,4-addition (Figure 4).

Polyisoprene stereoisomers with 1,2- or 3,4-structure have not yet been obtained in a sterically pure enough state to form crystalline products [73].

Conjugated dienes, unlike α-olefins, could be polymerized with homogeneous catalyst systems because of their electron-repelling conjugated double bonds which are easily coordinated [76]. All four possible stereoisomers of butadiene have been prepared with homogeneous catalysts. It should be noted that a high degree of steric purity could only be obtained with this catalyst system [89, 90].

The structure of these polymers has been studied by many researchers [57, 72, 76], especially by means of IR spectroscopy and X-ray analysis.

The synthesis of macromolecular compounds with asymmetrical structural units was finally achieved only recently, after the discovery of stereospecific polymerization.

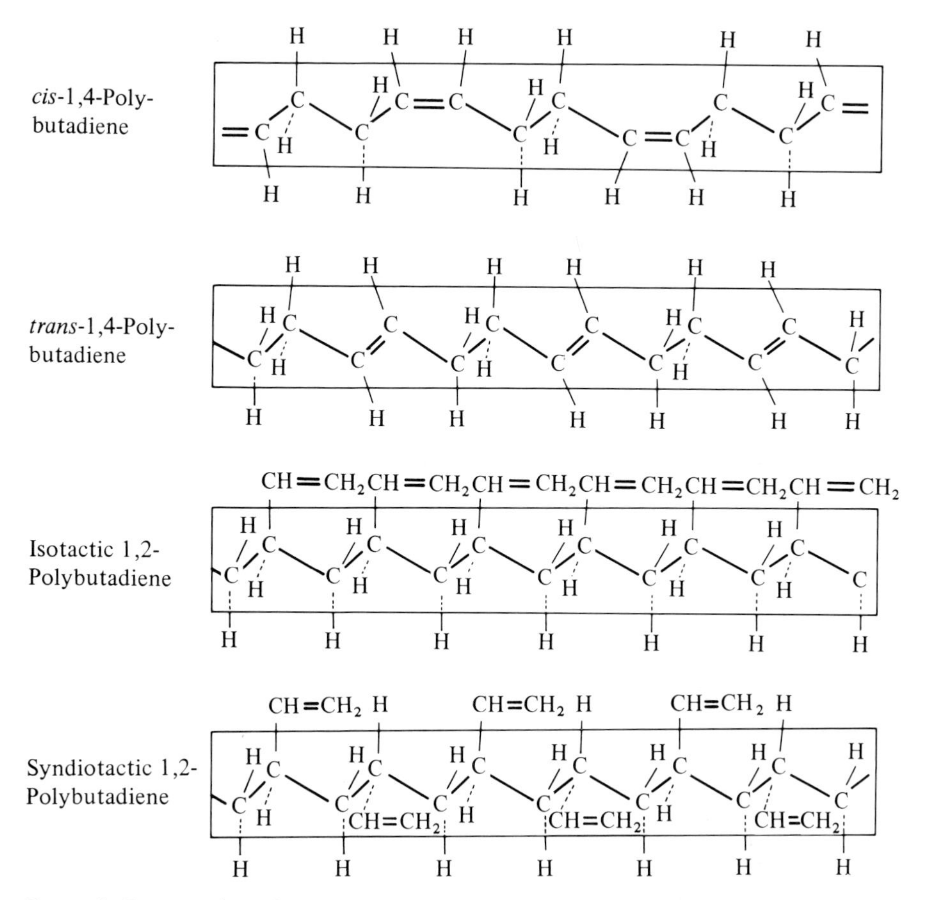

FIGURE 3. Stereoregular polybutadienes [88].

Natta [91] was able to prepare polymers of high optical purity starting from monomers without asymmetrical carbon atoms, but whose structure allowed the formation of two enantiomeric forms. Substituted butadienes were thus polymerized yielding tritactic *trans*-1,4-*erythro*-diisotactic polymers.

A vast new research field, thoroughly investigated during recent years, is the polymerization of monomers with different functional groups. Much information was gathered, contributing to a better understanding of the ionic or ionic-coordinative features of stereospecific polymerization mechanisms by studying those polymerization processes. Such polymerizations take place in homogeneous systems; the so-called "non-hydrocarbon" monomers, as Natta called them [73], contain functional groups with an unshared electron pair (ether, carbonyl or carboxyl oxygen atoms or amino, amido or nitrile nitrogen atoms). The stereospecificity is related to the uniform orientation of the

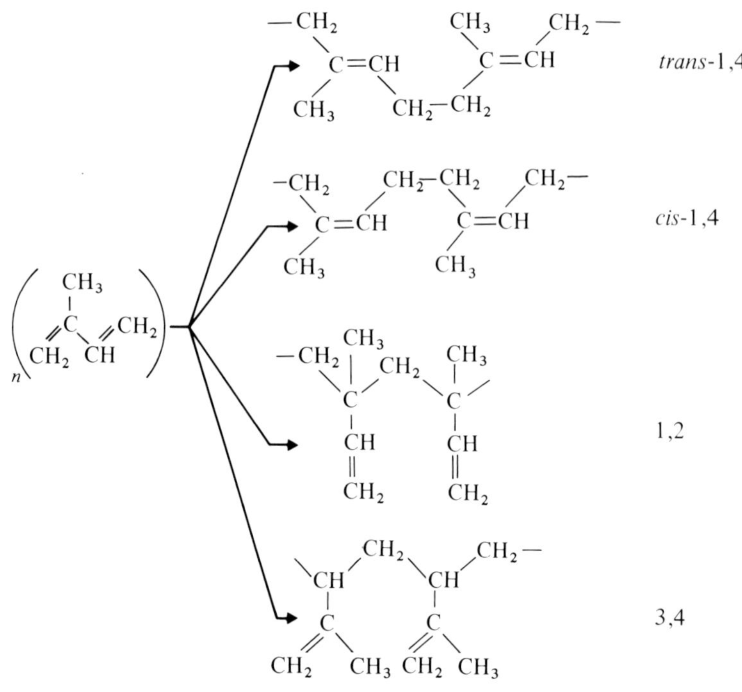

FIGURE 4. Stereoregular polyisoprene [86].

incoming monomer molecules with respect to the growing chain because such monomers are doubly attached to the catalyst complex by the olefin bond (obviously linked to the active site), and by the group at the end of the living chain.

Chapter II

Stereospecific catalysts

CATALYSTS play a highly important part in stereospecific polymerizations. Their action depends upon the monomer, the solvent and the polymerization conditions. The catalyst forces the monomer to assume a certain orientation during polymerization, thus ensuring the formation of long sequences or even macromolecules with identical units.

The classification of stereospecific catalysts according to the phase system in which polymerization occurs is discussed in the next sections. In fact, this classification is based on the physical state of the catalyst in the system studied.

A. Heterogeneous catalyst systems

In this section the following types of catalysts may be included based on more or less comparable criteria:

A_1. Ziegler–Natta-type catalysts

This catalyst class includes insoluble complexes prepared by mixing organometallic compounds of metals of groups I–III, Li, Al, Sn, Mg, Cd, Hg (in which the organic radical is usually aliphatic and only seldom aromatic), alkyl halides, especially aluminium derivatives, or their hydrides [67] with transition metal derivatives of groups IV–VIII such as halides, hypohalides, and oxyhalides of Ti, Zr, Ni, Ce, V, Mo, W, Co, etc. [18, 76, 92].

A characteristic feature of these catalysts is the dark precipitate formed on mixing the catalyst components. The molecular structure of these complexes is unknown, since it is difficult to isolate and to analyse the products. In such mixtures, derivatives of the lower valency state of the transition metals are predominantly formed [18, 93]. Some derivatives of the lower valency state of the transition metals could be prepared separately. By mixing them with metal alkyls, highly reactive products for olefin and diene polymerization were obtained [93–97]. Changes of the catalyst composition or even changes of the ratio of metal alkyl to transition metal derivative may produce changes of the mechanism [98, 99] or of the polymer microstructure [100].

The main factors influencing Ziegler–Natta-type catalyst activity which will be further discussed, are the following:

1. The nature of the active sites

The stereospecificity of such catalysts is limited by the concomitant presence in the mixture of various components. Heterogeneous systems, such as $AlR_3 + TiCl_3$ or VCl_3, have a great influence on the formation and stability of complex surfaces. A stable electronic configuration of the complex must be reached, involving a polarized carbon–metal bond, the organic moiety assuming the role of a carbanion.

Based on the existence of mixed metallic compounds, soluble in the reaction mixture and having a well-defined chemical structure, which are able to catalyse the polymerization process in a homogeneous system [80, 101, 102]:

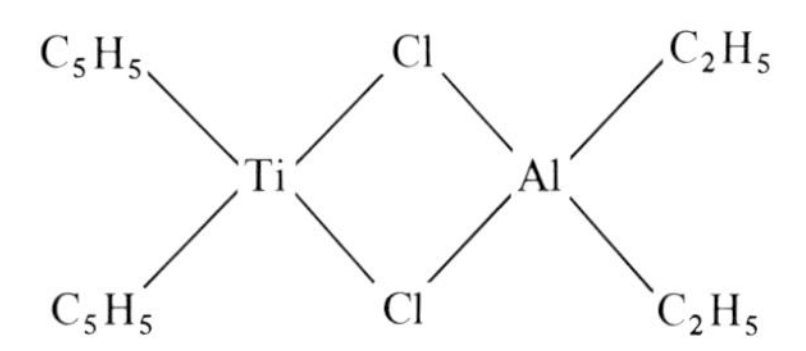

and also on the existence of aluminium or beryllium alkyls as dimers in benzene solution, the following structure was assigned to the catalyst complex:

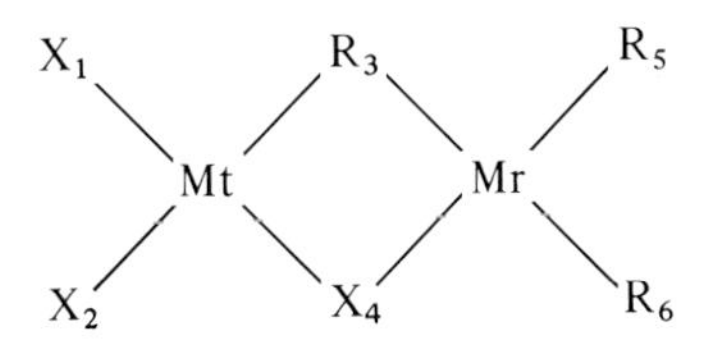

where Mt is the transition metal and Mr is the strong electropositive metal to which the alkyl groups are attached. The complex is adsorbed on the crystalline surface of the transition metal halide and in this state it initiates the stereospecific polymerization.

Titanium alkylation and reduction by the organometallic compound is a critical step in the polymerization mechanism. For instance, the formation of an organotitanium complex of the type $RTiCl_3$–$AlRCl_2$ has been suggested [103].

Uelzmann [104–106] and Bestian [107] suggest an ionic structure for the active site, such as $(TiCl_3)^+ (AlR_3Cl)^-$, where the negative complex ion creates the active polymerization site on the catalyst surface.

Olefins are able to form a stable compound with a transition metal [108].

By assuming that the crystal surface contains lattice defects, namely vacant atomic sites, the creation of active sites may occur via the alkylation of pentacoordinated titanium atoms [109–112].

$$Cl_5Ti{-}\square + Al(C_2H_5)_3 \longrightarrow Cl_5Ti{\cdots}C_2H_5{-}Al(C_2H_5)_2 \longrightarrow Cl_4Ti(\square){-}C_2H_5 + Al(C_2H_5)_2Cl$$

The coordination of the monomer to the transition metal (Figure 5) is a reaction occurring prior to the propagation step in Ziegler-type polymerizations [10, 111, 113–122]. For active centres, Cossee [109] proposes the configuration shown in Figure 6.

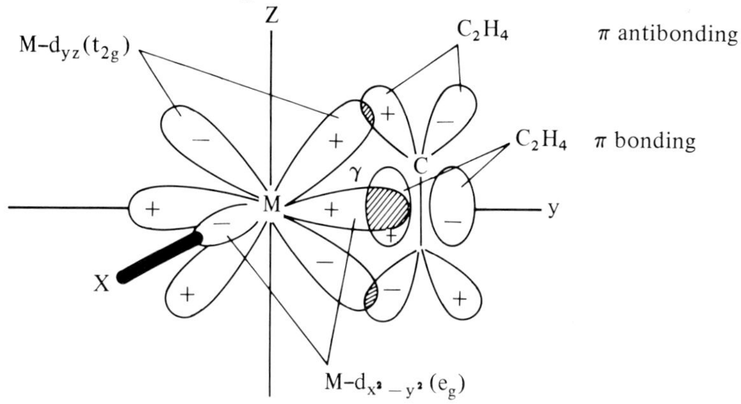

FIGURE 5. Spatial arrangement of orbitals in a π bond between a transition metal and ethylene [109].

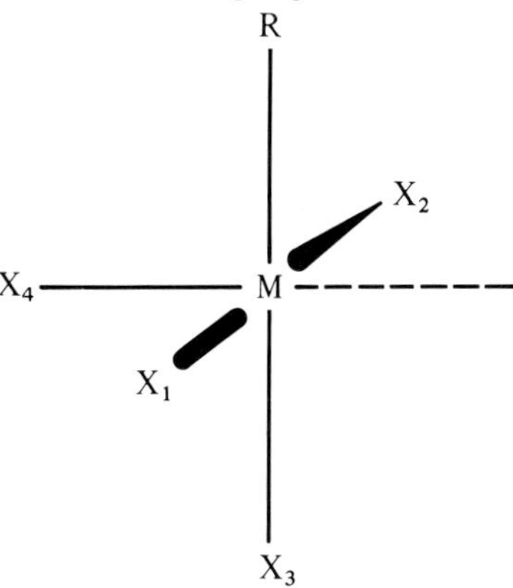

FIGURE 6. Assumed active site configuration of Ziegler–Natta-type catalyst systems. M = transition metal ion; R = alkyl (growing polymer chain); X_1–X_4 = anions [109].

The stereospecific polymerizations performed in the presence of transition metal halides and certain compounds containing no metal–carbon bonds [119, 123–124] or in the presence of alkyltitanium halides [102] confirm the existence of monometallic active sites.

The role of Lewis bases in Ziegler–Natta-type catalyst systems like $TiCl_3 + Al(C_2H_5)Cl_2$ is to produce the disproportionation of $Al(C_2H_5)Cl_2$, yielding $Al(C_2H_5)_2Cl$ which further forms with $TiCl_3$ the active catalyst complex [125]. Electron donors coordinate and inactivate all active sites apart from those showing a high stereospecificity [126]. The ability of electron donor compounds to form complexes on the catalyst surface is determined by their structure and by steric factors [127–131].

2. *The effect of the organometallic compound*

Since association phenomena occur, the reactivity of organometallic derivatives depends on the type of alkyl group and decreases in the order:

$$(CH_3)_3Al > (C_2H_5)_3Al > (C_4H_9)_3Al$$

The alkylating ability changes, however, in the reverse order [132]. Substitution of an alkyl group of the aluminium alkyls by a halogen atom reduces the reactivity in the following order [133]:

$$Al(C_2H_5)_3 > Al(C_2H_5)_2Cl > Al(C_2H_5)Cl_2$$

The reactivity of organometallic compounds increases with the increasing electronegativity of the reducing metal:

$$(C_6H_5)_3B > (C_2H_5)_3Al > (C_6H_5)_2Be > (C_6H_5)_2Mg > C_6H_5Li.$$

The smaller the ionic radius of the reducing metal ($Be < Al < Mg$), the higher the stereospecificity.

The molecular weight of the polymer as well as its stereospecificity are also influenced by the nature of the substituents [133–136].

3. *The effect of the transition metal compound*

Both the nature of the substituents and the valency state of the transition metal influence the catalyst activity [132, 137, 138]. For the polymerization of ethylene, the activity decreases in the following order:

$$TiCl_4 > C_5H_5TiCl_3 \geqslant (C_5H_5)_2TiCl_2.$$

The various crystalline forms of the transition metal compound determine the stereospecificity and the number of active sites; the changes in catalytic activity are related to the packing pattern of the atomic layers [93–97, 126, 139, 140].

The active sites, formed by alkylation of the titanium ions at sites where there are deficiencies in the chlorine atoms on the solid surface [109, 113, 141] have the necessary asymmetry for monomer orientation in order to form a complex with titanium [109–111, 142, 143] (Figure 7).

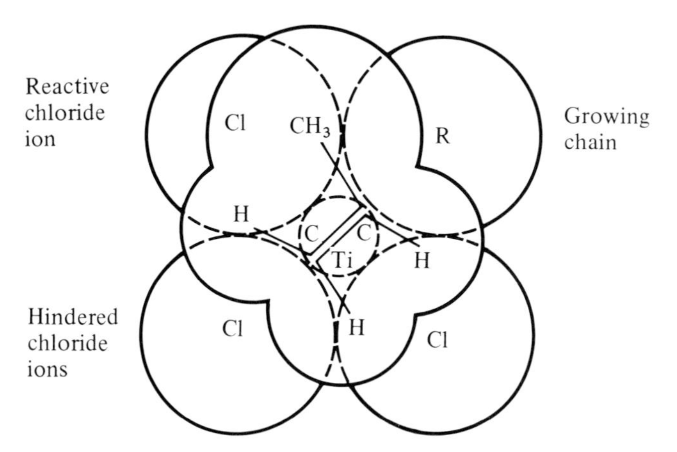

FIGURE 7. Relative position of propylene on an active site. Propylene molecule is in a perpendicular projection on the $TiCl_3R$ plane [111].

Topchev [144–146] assumes an exchange reaction between $TiCl_3$ and the adsorbed organometallic derivatives. The organotitanium compound further decomposes, forming a radical still adsorbed on the surface. This radical further initiates the polymerization of the monomer also adsorbed on the $TiCl_3$ surface.

4. The degree of dispersion and the physical state

Polymer stereoregularity is also dependent on the degree of dispersion of the catalyst and on its physical state. Complexes containing large particles yield preferentially isotactic polymers in the case of α-olefins, while finely dispersed particles suspended in the liquid system yield amorphous polymers [147, 148]. Experiments using separately the solid and the liquid products formed during the preparation of the catalyst complex, gave no conclusive results [76, 147, 149].

A_2. Oxide-type catalysts

Such catalyst systems are effective in the preparation of linear polyethylene and stereoregular polyolefins at low pressure, only a few atmospheres being required. Several types are known:

(1) chromium oxides supported on silica-alumina [150–152];
(2) molybdenum oxides, vanadium oxides or tungsten oxides supported on alumina [153];
(3) cobalt oxides or nickel oxides supported on activated carbon or on silica [154, 155];
(4) methods using promoters [156], in fact a variant of type 2 catalyst system. As promoters, alkali metals [157], alkali earth metals [158], group I–III metal hydrides or carbides [159–163], aluminium alkyls or alkali metal alkyls [162], etc., are used. To become effective, these oxide catalysts must be reduced in order to behave similarly to the reduced surface of the metal halide of Ziegler-type catalysts [69].

Cooper [138] suggested two structures for the active sites:

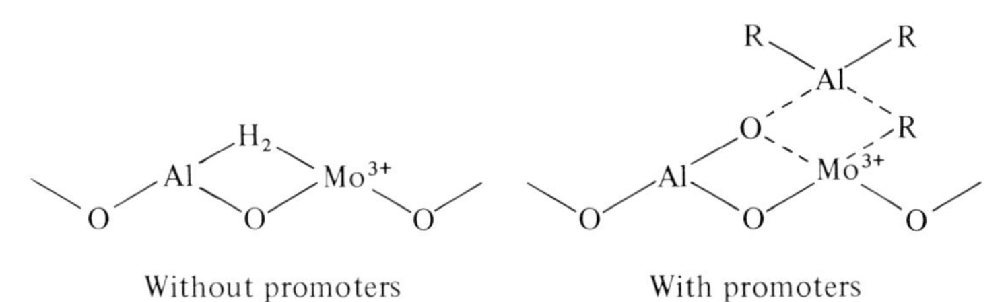

Without promoters — With promoters

A_3. Alfin catalysts

Alfin catalysts are a particular type of organometallic catalysts used more especially for butadiene polymerization; they are insoluble in hydrocarbon solvents, and consist of an alkali metal halide (NaCl), an alkali metal alkoxide (sodium isopropoxide) and a compound such as allylsodium [22, 164–167]:

$$C_5H_{11}Cl + 2Na \rightarrow C_5H_{11}Ns + NaCl$$

$$C_5H_{11}Na + (CH_3)_2CHOH \rightarrow (CH_3)_2CHONa + C_5H_{12}$$

$$C_5H_{11}Na + CH_2{=}CHCH_3 \rightarrow CH_2{=}CHCH_2Na \mp C_5H_{12}$$

A_4. Alkali metal, alkyl or aryl catalysts

Styrene has been polymerized to an isotactic polymer in hydrocarbon solvents in the presence of *n*-amyl sodium [168] or rubidium alkyls [169]. The reaction proceeds stereospecifically only in the presence of non-solvents for both the catalyst and the polymer. This fact proves the heterogeneous nature of the system, which according to Szwarc [170], is the key factor in its stereospecificity. Low temperatures are required to incorporate the monomer into a complex stable enough to control the formation of the polymer chain.

B. Homogeneous catalyst systems

Stereospecific polymerization may also occur in the presence of catalyst systems soluble in the reaction media [147, 171, 172].

Such systems are well suited for kinetic studies. In certain cases the catalysts could be isolated and analysed. The active catalytic sites correspond to the number of transition metal atoms [173].

The most used catalysts of this type are the following:

B_1. Organic compounds of group I–III metals

1. Lithium alkyls and lithium metal

Lithium metal and organolithium compounds are used in the polymerization of some vinyl and diene monomers [174–177].

Organolithium catalysts have a high coordination ability [178]. A monomeric lithium alkyl is able to complex both double bonds of a diene, providing the correct orientation for the formation of polymers with 1,4-structures.

$$\underset{\substack{\text{covalent}\\ \text{(I)}}}{\text{LiR}} \rightleftarrows \underset{\substack{\text{contact}\\ \text{ion-pair}\\ \text{(II)}}}{\text{Li}^+\,\text{R}^-} \rightleftarrows \underset{\substack{\text{solvent}\\ \text{separated}\\ \text{ion-pair}\\ \text{(III)}}}{\text{Li}^+|\text{R}^-} \rightleftarrows \underset{\substack{\text{free ions}\\ \text{(IV)}}}{\text{Li}^+ + \text{R}^-}$$

In hydrocarbon solvents, lithium alkyls behave as I and II, starting the polymerization process at a low rate, with a more marked steric effect than structures III and IV [178].

Alkyllithium compounds are present in hydrocarbon solvents as hexamers [179–189], or tetramers [190, 191], which dissociate according to the following equation:

$$(\text{RLi})_6 \rightleftarrows (\text{RLi})_4 + (\text{RLi})_2$$

The degree of association changes with the nature of the alkyl group [192]. By adding RLi to an olefin, a new organolithium compound results, affecting the kinetics of the reaction by association with the excess catalyst. The active species in diene polymerization is isoprenyllithium and butadienyllithium, respectively, both formed during the initiation step [182, 183, 192, 193]. Figure 8 shows the formation of this intermediate initiating species [183].

Lithium metal behaves similarly to the lithium alkyls. A metallation

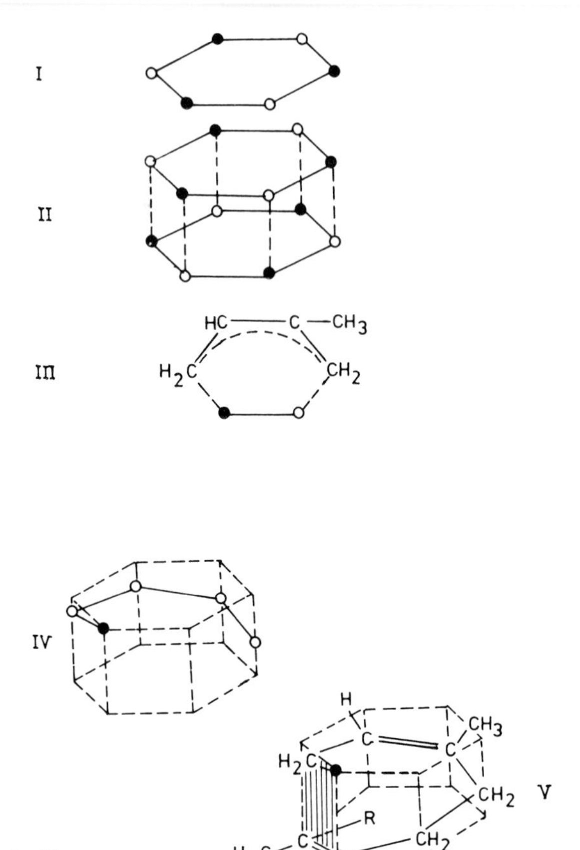

FIGURE 8. Structural models for associated forms of organolithium compounds: trimer (I) and hexamer (II), organolithium-isoprene adduct (III), unimeric butyllithium (IV), unimeric self-solvated polyisoprenyllithium (V) [183].

reaction of the monomer takes place [69, 103, 176, 194, 195] which, in the case of isoprene, results in the formation of isoprenyllithium.

$$2Li + CH_2{=}\underset{\displaystyle CH_3}{\underset{|}{C}}{-}CH{=}CH_2 \rightarrow LiCH_2{-}\underset{\displaystyle CH_3}{\underset{|}{C}}{=}CH{-}CH_2Li$$

2. Organometallic compounds of group II metals

Such compounds have been studied less. According to Kennedy [196] their activity and stereospecificity would be similar to the Na and K catalysts and are attributed to the ionic character of the metal–carbon bond and to the electropositivity of the metal.

3. Organometallic compounds of group III metals

Such compounds, mostly aluminium derivatives, have been used to prepare stereoregular polymers of styrene and other more polar monomers [197, 198]. Their stereospecificity was thought to be due to the formation of complexes, such as the following:

$$R_2Al(\mu\text{-}R)_2AlR_2 \rightleftharpoons R_2Al^{\oplus} + AlR_4^{\ominus}$$

B_2. Soluble Ziegler-type catalysts

Cobalt catalysts, used mostly to prepare polybutadiene with a high content (more than 99%) of *cis*-1,4 addition products, are the most widely known Aluminium alkyls or aluminium alkyl sesquichloride are used as co-catalysts [199–212].

The nature of active sites may be discussed in the same way as for the classical Ziegler-type catalysts.

Longiave [199] and Cooper [173] have observed that, independent of the anion present in the cobalt compound, the stereospecificity is always the same, a polymer with over 95% of *cis*-1,4 units being formed.

The first step in the synthesis of such complexes is the formation of $CoCl_2$:

$$>Al(Cl) + (RO)Co< \longrightarrow >Al(OR) + (Cl)Co<$$

followed by an alkylation reaction, leading to the unstable

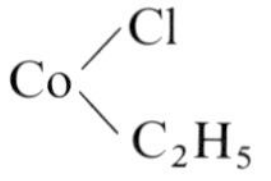

$$Co\begin{matrix} \diagup Cl \\ \diagdown C_2H_5 \end{matrix}$$

intermediate, which is probably converted to Co—Cl by a homolytic cleavage of the Co—C bond. It seems that this compound is stabilized in the monovalent state after complexing with $Al(C_2H_5)_2Cl$, forming in this way the active catalytic intermediate [199].

The monovalent state of the active catalytic intermediate was also suggested by van de Kamp [213] for the π-complex of cobalt with butadiene:

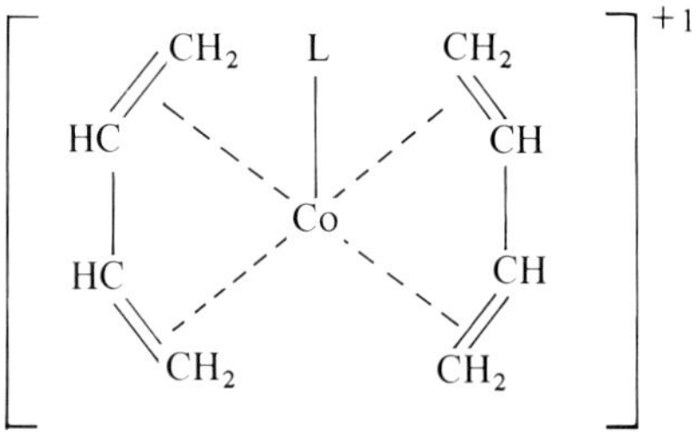

$$\left[\begin{matrix} & CH_2 & L & CH_2 & \\ HC & & | & & CH \\ | & & Co & & | \\ HC & & & & CH \\ & CH_2 & & CH_2 & \end{matrix} \right]^{+1}$$

where L is a Lewis base [213].

cis-1,4-Polybutadiene was prepared with cobalt derivatives complexed with compounds of the following type:

$$R_2AlCl,\ RAlCl_2,\ RClAl{-}O{-}AlClR,\ RClAl{-}NR{-}AlClR$$

The substituted aluminium alkyls supply the chlorine atoms, the presence of which seems to be essential for the formation of the active sites [199, 214].

The soluble catalyst systems based on cobalt salts and $AlCl_3$ are also effective to prepare *cis*-1,4-polybutadiene.

Scott [215] made use of the $CoCl_2 + AlCl_3$ system. Its activity is independent of the Al/Co ratio and becomes stereospecific only in the presence of thiophene or Al, Mg or Zn powders.

For the polymerization of butadiene, catalyst systems with nickel as the transition metal [216], in the presence of other metal halides necessary to form halogen bridges within the complex [217–221], have also been used.

Compounds containing chromium, as the transition metal, yield stereoregular 1,2-polybutadiene, when chromium acetylacetonate and aluminium alkyl were used [214].

Soluble compounds resulted from the reaction of an alkyl aluminium with titanates, such as $AlR_3 + Ti(OR)_4$, have also been used for the polymerization of ethylene and butadiene [212–225].

Bawn and Symcox [223] proved that in the polymerization of ethylene with $Al(C_2H_5)_3 + Ti(OC_4H_9)_4$ the active site is a soluble bimetallic compound having the following structure:

```
                CH3
                 |
RO              CH2              C2H5
   \          /     .          /
     Ti              Al
   /          .     /          \
RO              CH2              C2H5
                 |
                CH3
```

C. Catalysts effective in polar media

Some monomers, such as butadiene, isoprene and cyclobutene, have been converted into stereoregular polymers by polymerization in strong polar media (aqueous emulsion, ethyl alcohol, dimethylformamide) [226–230]. The catalysts used are salts, mainly chlorides, of group VIII metals, namely Rh, Ru and Ir.

The actual structure of the active sites was not elucidated. As for Ziegler-type catalysts, the formation of a metal–carbon bond was assumed, with the concomitant evolution of hydrogen chloride [231]:

$$\text{ru—Cl} + \begin{array}{c}\text{HC=CH}\\ |\quad\;\, |\\ \text{H}_2\text{C—CH}_2\end{array} \rightarrow \text{ru—CH=CH—CH=CH}_2 + \text{HCl}$$

Dawans and Teyssié [136], on the basis of some rhodium complexes with certain dienes, such as 1,5-cyclooctadiene, isolated by Chatt [232], to which a dimeric structure was assigned

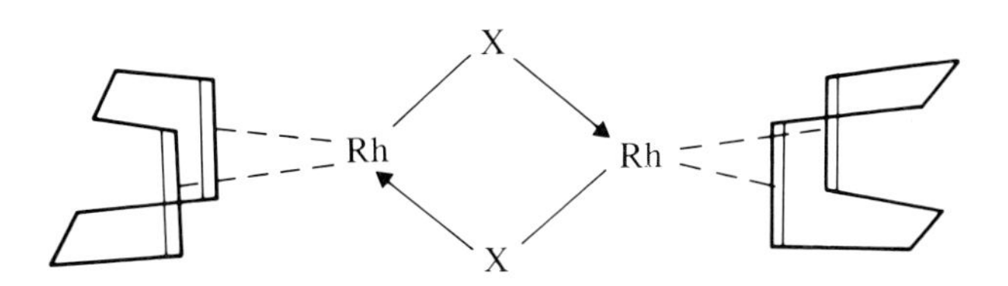

made an analogy between the activity of these complexes and the active sites in Ziegler-type catalysts:

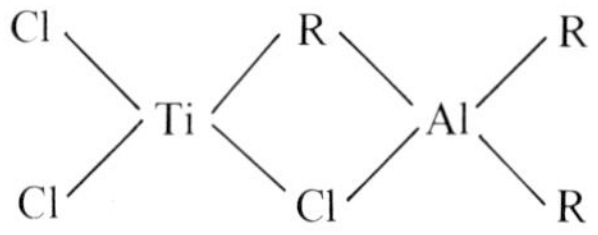

$$\begin{array}{ccccccc} \mathrm{Cl} & & & \mathrm{R} & & & \mathrm{R} \\ & \diagdown & & & & & \\ & & \mathrm{Ti} & & \mathrm{Al} & & \\ & \diagup & & & & & \\ \mathrm{Cl} & & & \mathrm{Cl} & & & \mathrm{R} \end{array}$$

where the titanium atom plays the role of a noble metal. The addition of small amounts of cyclohexadiene to the reaction mixture produces a cyclohexadiene–rhodium complex, stabilizing rhodium in its valency state of one [233].

Polymerizations with $RhCl_3$ in aqueous emulsion occur only on the surface of the complex; the formation of the active complex is dependent on the type and on the length of the alkyl group of the surface-active agent. When alcohol is used as solvent, polymerization occurs in a homogeneous system and no interfacial conditions are required [234].

Chapter III

Mechanisms of stereospecific polymerization

SEVERAL papers have been published dealing with the mechanism of stereospecific polymerization. Such polymers have a particular structure from the point of view of both the sequence of the ordered atoms and the spatial configuration.

Although first approaches to the mechanism of stereospecific polymerizations attributed the main role to the surface of the catalyst and its ability to adsorb and orientate the monomer, more recent investigations state that polymerizations initiated by free radicals under controlled reaction conditions, can yield stereoregular polymers starting from monomers with more polar substituents, by using external factors markedly favouring a given transition state [72, 76, 80, 235, 236].

Stereospecific polymerization requires one of the following conditions [76, 86, 237]:

— macroscopic surfaces (SiO_2, Al_2O_3, carbon) able to adsorb or complex the ion pair and in this way able to direct the reaction towards the formation of a stereoregular polymer (Phillips-type catalysts);

— microscopic surfaces as in the very finely dispersed suspensions of Ziegler-type catalysts;

— the resulting polymer molecule may further behave as a matrix for the incoming monomer. On solution the polymer induces a certain order in the liquid reaction medium and forces the monomer into a suitable orientation for the formation of a linear polymer, by adsorption on the polymer already formed.

— bulky counter-ions inducing a preorientation of the monomer; complex formation of the catalyst with monomers takes place leading to the formation of a transition state stable enough to cause the addition of the monomer to the growing chain even in homogeneous solutions.

The classification of chain polymerizations into free radical and ionic takes into account the nature of the intermediate species involved in chain propagation (macro-radicals or macro-ions) and the mechanism of formation of these intermediates.

In a free radical mechanism, the reaction occurs via free radicals generated by photochemical or thermal methods or with initiators or high energy radiations. In this case a homolytic cleavage of the double bond takes place:

$$R\cdot + \overset{|}{\underset{|}{C}}=\overset{|}{\underset{|}{C}} \rightarrow R-\overset{|}{\underset{|}{C}}-\overset{|}{\underset{|}{C}}\cdot \quad \text{free radical}$$

Ionic polymerization is governed by thermodynamic and chemical requirements. The active sites are ions; two cases are possible:

$$X^- + \overset{1}{C}H_2=\overset{2}{C}H(R) \rightarrow X-\overset{1}{C}H_2-\overset{2}{C}H^-(R)$$

carbanion

The resulting carbanion initiates the polymerization through anionic chains.

If the growth of the macromolecule occurs through carbonium ions, the initiation of the polymerization takes place through cationic chains.

$$Y^+ + \overset{1}{C}H_2=\overset{2}{C}H(R) \rightarrow Y-\overset{1}{C}H_2-\overset{2}{C}H^+(R)$$

carbonium ion

In each case, the product is an active species similar to the initiating species and each new step regenerates the active species.

The free radicals could be detected through various methods, the ionic intermediates, however, could not be detected in a free state. In media with various dielectric constants, they are more or less associated with the counter-ion.

One and the same catalyst may polymerize certain monomers by either a free radical, ionic or ionic coordination mechanisms. It is not yet possible to predict the behaviour of various monomers toward the numerous catalyst systems.

Furukawa [238] assumed that the reactivity of vinyl monomers is controlled by their polarization and their stabilization through resonance. From the resonance stabilization (Q) and the polarity of a vinyl monomer molecule (e) he found a quantitative relationship for monomer reactivity. He states that the higher the *e* value of a given monomer, the higher its polymerizability via an anionic mechanism; conversely, at low *e*-values, a cationic mechanism is most probable.

A relationship between the reduction potential (measured by polarography) and the polymerizability of a vinyl monomer via a particular mechanism was also established. Monomers with low reduction potentials preferentially undergo polymerization by anionic mechanism.

Another general monomer classification, correlating their polymerizability with a given polymerization mechanism was suggested by Sigwalt [239]:

— Almost all monomers can undergo a free radical homopolymerization; the influence of the polarity of the double bond substituent is less important for such mechanisms.

— Monomers containing strong electron-repelling groups can be polymerized by a cationic mechanism e.g.:

$CH_2 = C(CH_3)_2$ (Isobutylene), Indene or Cyclopentadiene [ring structures with CH_2]

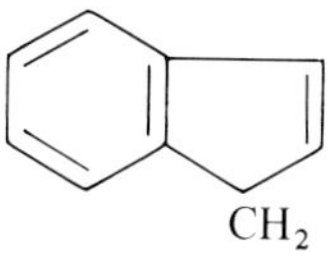

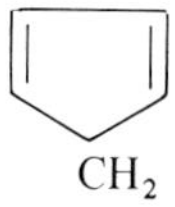

Isobutylene Indene Cyclopentadiene

— Monomers containing electron-attracting groups facilitating the attack of an anion on the double bond can be polymerized by anionic mechanisms e.g.:

$$CH_2{=}CCl_2$$

vinylidene chloride

Monomers, such as

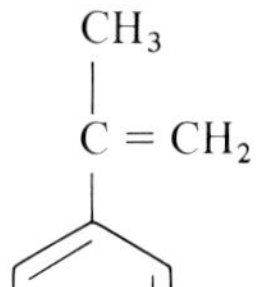

$$C_6H_5{-}C(CH_3){=}CH_2$$

α-methyl-styrene

are more likely to polymerize by a cationic mechanism, while monomers such as

$CH_2{=}C{-}CH{=}CH_2$ with CH_3 substituent on C — Isoprene

$CH{=}CH_2$ attached to a benzene ring — Styrene

$CH_2{=}CH{-}CH{=}CH_2$ — Butadiene

Isoprene Styrene Butadiene

polymerize by an anionic mechanism.

A particular class includes ethylene monomers easily polymerized with complex catalysts.

The mechanism of this latter type of polymerization (although controversial) shows most of the general features of ionic polymerization. According to Kennedy [196], "simple" polymerizations occur through conventional free radical and ionic polymerization mechanisms when the propagating species is free and not influenced by initiators or counter-ions, usually yielding syndiotactic polymers. The orientation forces of the monomer are weak and polymers with low tacticity are formed. Polymers with a high degree of tacticity may be obtained only from certain monomers at low temperatures.

When other forces, apart from the catalyst and the polymer chain end groups, are also involved in the monomer orientation, tacticity can be increased (channel-type complexes, solid-state polymerizations).

In the following sections, the various polymerization mechanisms will be discussed:

A — free radical mechanisms;
B — ionic mechanisms;
B_1 — cationic mechanisms:

1 — classical cationic mechanisms;
2 — cationic coordination mechanisms;

B_2 — anionic mechanisms:

1 — classical anionic mechanisms;
2 — anionic coordination mechanisms;

C — mechanisms involving Ziegler–Natta catalyst systems;
C_1 — free radical coordination mechanisms;
C_2 — cationic coordination mechanisms;
C_3 — anionic coordination mechanisms.

A. Free radical mechanisms

Polymerizations initiated by free radical mechanisms have been thoroughly studied. However, the probability of forming stereoregular polymers by such polymerization mechanisms is very low. Fox and co-workers [240] were the first to try to prove that such processes were sterically and thermodynamically possible.

The use of appropriate temperatures and reaction conditions converts vinyl monomers into stereoregular polymers in a homogeneous system even when free radicals are involved. The lack of marked stereoselectivity for a head-to-tail polymerization and random additions reduce the polymer stereoregularity.

A reduction of the temperature increases the stereospecificity and, implicitly, the crystallinity [21, 73]. Among the thermodynamic factors, the low differences of energy between the isotactic and the syndiotactic growing steps must be also taken into account. At normal temperatures, both configurations result except when R is very bulky.

The steric orientation is, however, very limited and high tacticity is seldom realized by this mechanism (for instance the stereospecific polymerization of some non-hydrocarbon monomers, such as methyl methacrylate). This monomer polymerized with free radical initiators at low temperatures yields a crystallizable polymer.

Stereospecificity arises from the free energy differences of two transition states produced by interaction of the ultimate and penultimate monomer units in each monomer addition step [79]. Therefore, stereospecific polymerization by a free radical mechanism can occur only when a selective orientation of the monomer addition is possible depending on the type of monomer and, especially, on its substituents. In a vinyl monomer, $CH_2{=}CHR$, the R group attached to the carbon atom involved in the active site, $-\dot{C}H-R$, can favour an alternate (syndiotactic) arrangement since it is the less hindered position for the R group.

The synthesis of *trans*-1,4-polychloroprene by free radical polymerization is possible probably on account of the electronic dissymmetry induced by the chlorine atom [173]. When the stereoregularity is controlled only by the steric effects of the monomer substituents, the mechanism of stereospecific polymerization is similar to that of free radical polymerizations and syndiotactic polymers are formed.

Also in the case of dienes, there are differences between the activation energies of the various addition modes [241].

According to Kern [242], syndiotactic polymers can be formed in a homogeneous medium by free radical polymerization, if the specific interactions between substituents do not favour an isotactic configuration. No isotactic polymers can be prepared by free radical polymerization in homogeneous media.

The initiators acting in a homogeneous medium by a free radical mechanism are not involved in the chain-growing step. Hence, in diene polymerization, the only factor controlling the configuration is the *cis* or *trans* structure of the diene. The temperature is also critical since it controls the monomer *cis–trans* equilibrium [242].

In free radical stereospecific polymerization no stereoregulating surface is involved and only syndiotactic polymers with a low degree of tacticity can be formed [196].

B. Ionic mechanisms

The higher complexity of ionic polymerizations, as compared with free radical polymerizations, is explained by the following facts:

— the active species is usually a polar complex and the counter-ion influences the reaction course;

— the large interaction between the active species and the reaction medium generated by their polar structure;

— there are effects provided by ion dissociation or agglomeration.

The major role played by the counter-ion in stereospecific polymerizations is to orientate the monomer incorporation into the growing chain [239].

Sigwalt [239] and Danusso [243] have shown that the growth step can take place in different ways according to the various interactions of the monomer with the ions present in the reaction mixture.

In the "classical ionic" systems, the monomer is directly attached to the carbanion or to the carbonium ion (ions can be free or associated in ion pairs).

— The monomer is attacked by a carbanion and is inserted between the two ions

$$\underset{\substack{\downarrow \\ \mathrm{M}}}{\text{——}(-)}(+) \rightarrow \text{——}\mathrm{M}(-)(+)$$

This reaction is favoured by electron-attracting substituents at the double bond of the monomer.

— The monomer is attacked by a carbonium ion and then inserted between the two ions:

$$\underset{\substack{\downarrow \\ \mathrm{M}}}{\text{——}(+)}(-) \rightarrow \text{——}\mathrm{M}(+)(-)$$

Electron-repelling substituents at the double bond favour this reaction.

The ionic coordination systems where fixation at the carbanion or at the carbonium ion is preceded by coordination of the monomer with the counter-ion:

— The monomer is attacked by a cation (the carbanion counter-ion) and remains attached to it:

$$
\text{——}(-)(+)\underset{\substack{\downarrow\\ \text{M}}}{\ }\rightarrow \text{——M}(-)(+)
$$

The reaction is favoured by electron-repelling substituents.

— The monomer is attacked by an anion (the carbonium ion counter-ion) at which it remains attached:

$$
\text{——}(+)(-)\underset{\substack{\downarrow\\ \text{M}}}{\ }\rightarrow \text{——M}(+)(-)
$$

The reaction is favoured by electron-attracting substituents.

B_1. Cationic mechanisms

Any compound able to convert a double bond into a carbonium ion is a virtual catalyst for cationic polymerization.

Polymerization can occur only when thermodynamic and kinetic conditions allow the carbonium ion to react with other monomers. Olefins, α-olefins and 1- and 4-substituted-1,3-dienes can be polymerized via this mechanism. In such polymerizations, catalysts play an essential role [244].

The ionic species can initiate several concurrent reactions and, therefore, low temperatures favour cationic and especially stereoregular polymerization.

1. "Classical" cationic mechanisms

Up to date, cationic polymerizations have been less studied and their importance for hydrocarbon monomer polymerization is limited. The stereospecific cationic polymerization of vinyl ethers in a homogeneous medium has been studied in detail [245]. Catalysts such as $(AlClR_2)_2$, $(AlCl_2R)_2$ or $TiCl_2(OR)_2$, able to form etherate complexes, have been used. Under such conditions, the monomer can be satisfactorily oriented within the complex and incorporated in a stereoregular configuration into the growing chain. The stereospecificity and stability of the complex are dependent on temperature, solvent and the choice of a suitable ion.

According to Kennedy [196] the catalysts which act by a simple cationic mechanism are of $MX_n^- R^+$ type. The catalyst—R bond being ionic, and the two ionic species fully separated, syndiotactic polymers with a low degree of tacticity are formed.

2. Cationic coordination mechanisms

According to Dawans and Teyssié's suggestions [136] on the mechanism of polymerization with rhodium catalysts in polar media, the cationic coordination polymerization mechanism is related to the formation of a relatively stable complex between the monomer and the metal atom. In this complex (Figure 9), the π double bond electrons form a σ bond with the metal *s* orbitals, while the *d* orbitals overlap with the antibonding orbitals of the double bond carbon atoms.

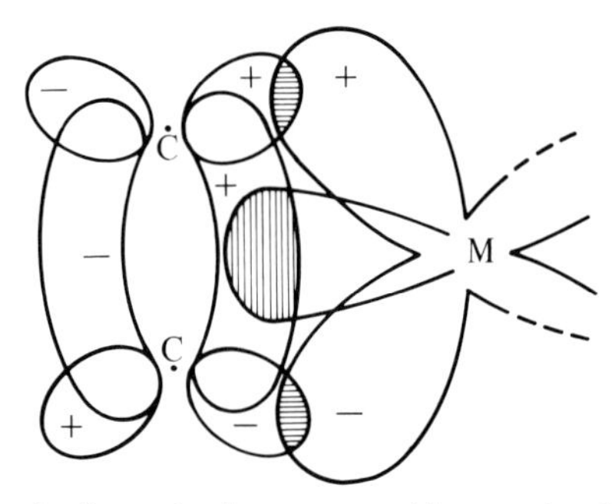

FIGURE 9. Scheme of complex formation between transition metal and double bond of monomer [136].

B_2. Anionic mechanisms

1. Classical anionic mechanisms

In simple anionic polymerizations, the growing anion is a free ion, the cation providing only electrical neutrality [196]. The above definition, however, must be limited, according to Duck [90], only to those catalysts having a high degree of ionization and saline properties, such as sodium alkyls [246–248], potassium alkyls [241–250] and caesium or rubidium alkyls. Lithium metal or organolithium catalysts, unlike sodium or potassium alkyls, only initiate simple anionic polymerization in basic solvents with monomers which form resonance-stabilized polymeric anions.

Alkali metal alkyls (sodium, potassium or rubidium) behave as stereospecific initiators in heterogeneous systems [168]. The organic group, namely alkyl, aryl or aralkyl, has a reduced influence on stereospecificity [251, 252].

Sodium alkyls in hydrocarbon solvents [169, 250, 253–255], Alfin catalysts or potassium alkyls [169, 250, 256] may yield isotactic polystyrene.

Solid surfaces and low temperatures are necessary for stereospecific orientation [168, 257]. Styrene polymerized with sodium alkyls in hydrocarbon solvents, at temperatures ranging from $-20°$ to $+30°C$, has about the same crystallinity as when Ziegler-type catalysts are used. Stereospecificity also increases with increasing ionic character of the metal – carbon bond and with increasing ability of the metal cation to complex the monomer, in the order $K < Na < Li$.

Lee and his co-workers [258] assume for styrene polymerization with sodium derivatives two successive steps:

— complex formation between the monomer and the catalyst;
— rearrangement of the complex.

$$\sim\!\sim S^-, Na^+ + S \rightleftarrows \sim\!\sim S^-Na^+(S) \longrightarrow \sim\!\sim SS^-Na^+$$

Brown and Kern [168] studied the stereospecific polymerization of styrene with amylsodium and other insoluble alkali alkyl initiators and concluded that:

— the polymerization is anionic, heterogeneous, stereoregularity being the result of chain growth on the solid surface; tacticity and implicitly the polymerization rate is improved by lowering the temperature;
— the initiation involves the formation of alkyl anions on the solid catalyst surface;
— most of the initiated polymer chains remain in an active state until deactivating agents are added.

Thus, the formation of "living" polymers in a heterogeneous medium was proved to be possible [259].

It seems that isoprene also forms such polymers [260–262]. O'Driscoll [263] assumes, for polymerization with lithium dispersions, that the initiation step is the transfer of an electron from the metal to the monomer.

Polymerizations initiated with metal alkyls in the presence of electron donor compounds, added in specified amounts to the reaction mixture, which increase the polymerization rate, but have an unfavourable effect on stereospecificity and on the degree of polymerization, can also be included in this class [264]. The behaviour of lithium alkyls is well known. With electron donors containing oxygen atoms (e.g. tetrahydrofuran), a complex ion results.

Polymerization of styrene, isoprene and butadiene with ethyllithium in strong electron donor solvents, such as tetrahydrofuran, alters the initiation mechanism of the catalyst. The polarization of the Li—C bonds is increased and the charges are almost separated. In basic solvents, lithium forms complexes including two base molecules [179, 265, 266]. Therefore only one orbital is available to coordinate the monomer and the Li—C bond is more ionic, both factors favouring 1,2- and 3,4-addition. This reaction mechanism is related to the "classical" anionic polymerization where the active sites are carbanions [264]. The addition of electron donors forming complexes with the lithium atom of the active site reduces its effect on the chain growth.

2. Anionic coordination mechanisms

In this class are included catalyst systems based on lithium compounds, which are soluble in hydrocarbon solvents and are, therefore, efficient in homogeneous media.

The behaviour of lithium alkyls differs from sodium alkyls. The former have a more marked covalent character, although they are partly ionized. The covalency of lithium in its organometallic compounds is still controversial.

Coates [267] emphasizes the marked degree of covalency, while Warhurst [268] assumes that the Li—C bond in metal alkyls is about 80 % ionic. The ionic character of this bond is dependent, to a certain extent, on the resonance stabilization of the alkyl or aryl anion. The lithium aromatic derivatives seem to be completely ionic [269].

The dielectric constant of the polymerization solvent also controls, to a lesser extent, the association of the ions. In benzene, the ion pair is associated, while in tetrahydrofuran, kinetic evidence proves that the ion pairs are free [270]. Tetrahydrofuran solvates the ions cleaving the ion pair bonding and associates preferentially with one of the two ions, thus changing the orientation of the monomer and, hence, the microstructure of the polymer.

Organolithium compounds are also effective in the stereospecific polymerization of dienes. It seems that in the polymerization with lithium metal, intermediate addition compounds between monomer and lithium are formed, further initiating homogeneous polymerization [103, 176, 194, 195].

With the butyl lithium–isoprene system a high stereospecificity results in solvents with low dielectric constant, where the Li^+ counter-ion was initially solvated by the lithium alkyl [271]. This result proves that the counter-ion plays an important role in stereoregularity.

In his research on isoprene polymerization with butyl lithium in hydrocarbon solvents, Kuntz [272] proved that the *cis*-1,4-coordination complex of isoprene is less stable than the corresponding *cis* or *trans* form of the butadiene complex (Figure 10) and therefore the temperature is critical for the final structure in the case of the isoprene–butyl lithium system.

The coordination complex between the catalyst and the diolefin sterically

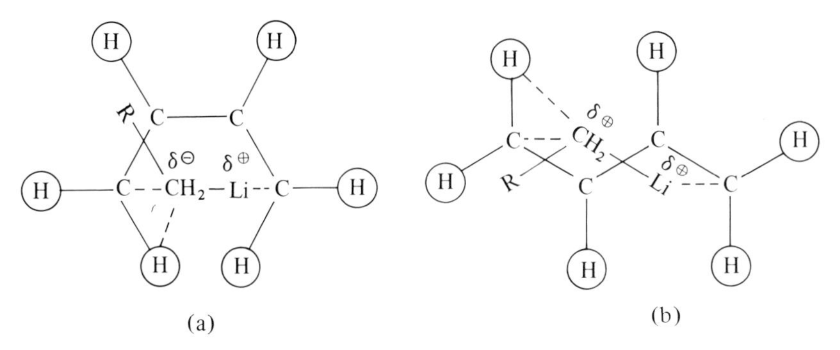

FIGURE 10. Schematic representation of the *cis* and *trans* complexes of butadiene with the growing polymer chain; (A) *cis*-complex; (B) *trans*-complex [272].

directs the inclusion of the nearest monomer unit into the growing polymer chain.

In the case of butadiene, the coordination complex includes a large number of *cis*-complexed molecules. The C—Li bond of the polymer must convert the butadiene molecules with a *trans*-configuration in solution into the *cis*-configuration of the molecules involved in the coordination complex.

The partly anionic carbon atom of the C—Li bond may coordinate either a hydrogen atom, or an electron-deficient carbon atom of the diene, or both of them. Therefore, stereospecificity arises during the simultaneous formation and cleavage of the monomer–complex bond. The C—Li bond of the growing polymer chain is in equilibrium with an aggregate-type structure including butadiene molecules and shows a rather low tendency to associate with other organolithium compounds. The growth of the polymeric chain occurs within such complexes.

C. Mechanisms of polymerization using Ziegler–Natta catalyst systems

Ziegler–Natta-type catalyst systems yield stereoregular polymers via anionic, cationic or even radical coordination mechanisms. With $TiCl_4$ in excess, the catalyst surface is predominantly cationic, while with AlR_3 in excess, owing to the adsorption of aluminium alkyl on the surface of the titanium compound, the catalyst surface is anionic [273]. The striking property of these catalyst systems is the monomer coordination onto the catalyst complex before the addition of the monomer to the reactive end of the partly stabilized growing polymer chain. Coordination occurs though the π-electrons or the unshared electron pairs of the monomer and the free orbitals of a metallic compound.

The end of the polymer chain is fixed in a given position and is partly stabilized by simple or complex counter-ions. The polymerization and orientation of the monomer must favour an isotactic placement [274]. Among the various possible transition states between monomer molecules and the growing chain end, the transition state involving the lowest activation energy will control the preferential addition of monomer molecules to the growing chain resulting in the formation of a stereoregular chain [27].

Isotacticity depends on the catalyst–polymer bond strength, the stability of the monomer–catalyst complex and on the stiffness of the surface achieving the orientation.

The "surface" can be a soluble asymmetric counter-ion or a solid surface, completely hindering the monomer approach from one side of the active site [274].

In order to proceed stereospecifically, the counter-ion–monomer complex must be more stable than the monomer–catalyst complex. This explains the behaviour of monomers yielding isotactic stereoregular polymers only in heterogeneous media (α-olefins).

C_1. Free radical coordination mechanisms

This type of mechanism is an extreme case of polymerization with Ziegler-type catalysts [196]. The catalyst has the general formula $MR \cdot MX_n$, with a covalent metal–carbon bond. The catalyst has a moderate electrophilic character, the stereoregulating surface is complex and a medium degree of tacticity is obtained. The free radical addition to the double bond is less selective in producing a head-to-tail structure of the polymer.

The tacticity is lower for α-olefins since the M—R covalent bond cannot polarize the monomer.

The free radical coordination mechanism is less usual since the most reactive organometallic compounds have an ionic character and the asymmetric monomers are more easily polarized by ionic catalysts.

Friedlander and Oita [275] have suggested a free radical mechanism for heterogeneous polymerization with the $AlR_3 + TiCl_4$ system; the free radical would result from a homolytic cleavage of an alkyltitanium halide. The radical adds to an olefin molecule resulting in the formation of a radical-ion:

$$M\cdot + CH_2 = CH_2 \rightarrow M{-}CH_2{-}CH_2\cdot \rightleftarrows M^{(+)(-)}CH_2{-}CH_2\cdot$$

The chain propagation takes place through this radical-ion, by reaction with the oriented monomer adsorbed on the catalyst surface. This surface is responsible for monomer orientation.

As the polymer chain is growing, the polymer is desorbed from the catalyst surface and replaced with the monomer, which orientates on the surface. The solvent facilitates the desorption of the polymer from the catalyst surface.

Duck [276] suggested a free radical mechanism for the polymerization of ethylene with the $Al(C_2H_5)_2Cl + TiCl_4$ system.

A free radical initiation is also suggested by Badin [236] for both the $TiCl_4 + Al(i\text{—}C_4H_9)_3$ and $TiCl_4 + LiC_6H_5$ systems:

$$TiCl_4 + AlR_3 \rightarrow RTiCl_3 + AlR_2Cl$$
$$RTiCl_3 \rightarrow TiCl_3 + R\cdot$$

C_2. Cationic coordination mechanisms

In the cationic coordination mechanism, the terminal atom of the growing chain is more electropositive than the counter-ion of the catalyst complex [133].

Monomers which more readily stabilize the carbonium ion can be polymerized by this mechanism.

Unlike the anionic coordination mechanism, the charge is situated in this case on an internal carbon atom or on the most substituted carbon atom of the double bond and not on the terminal carbon atom.

The cationic coordination polymerization process has been less studied and successfully applied only at low temperatures where chain transfers are avoided [15, 82]. In this mechanism the carbonium ion

$$(\geqslant C^+)$$

is attached to a strongly electronegative non-metallic atom (e.g. —O— or Cl—) belonging to the complex [277]:

$$>M\text{—}X + X\text{—}CHR\text{—}CH_2\cdots \rightarrow \left[>M\begin{matrix}\diagup X\\ \diagdown X\end{matrix}\right]^{(-)} \overset{(+)}{C}HR\text{—}CH_2\cdots$$

Various ions linked through coordinate bonds to the central atom of the complex tend to occupy the same position; therefore the X—C bond dissociation is more facile. The carbon atom can add to any coordinately bonded chloride ion of the complex. This fact, together with the large size of the counter-ion, explains the low degree of stereoregularity produced by this mechanism. The more substituted the carbonium ion, the more thermodynamically stable it is, favouring an isomerization reaction of the following type [277]:

$$\overset{(-)}{Cat}\text{—}\overset{(+)}{C}HR\text{—}CH_2\text{—}CHR\text{—}CH_2^- \rightarrow RCH_2\text{—}CH_2\text{—}\underset{\underset{(-)}{Cat}}{\overset{(+)}{C}R}\text{—}CH_2\text{—}$$

which produces irregularities in the chain structure. The forces keeping together the growing chain and the initiator are weaker as in anionic polymerization [278–279].

Sinn, Winter and Tirpitz [280] assume that polymerizations of styrene, isoprene, butadiene and dimethylbutadiene with $AlR_3 + TiCl_4$ or $AlRCl_2 + CoCl_2$ occur by a cationic coordination mechanism, since it takes place only in the presence of traces of water. This feature is specific for cationic polymerizations. The above authors have obtained tactic polystyrene by using the $TiCl_4 + AlR_2Cl + H_2O$ catalyst system.

Childers [279] suggests a cationic mechanism for butadiene polymerization with the $Al_2(C_2H_5)_3Cl_2$—Co octoate catalyst system yielding polymers with a high content of *cis*-1,4 structure. The cationic mechanism is made evident by using tritiated alcohols, $C_4H_9OH^*$, or C^{14}-labelled alcohols, $^*CH_3OH$, to stop the reaction.

C_3. Anionic coördination mechanisms

Anionic coordination polymerization is an addition process in which each monomeric unit is complexed and polarized onto a positive site, and the growing chain is an anion [139]. The catalyst is a complex, where usually a transition metal acts as coordinating agent; a carbon atom, involved in a growing polymer chain, is coordinated to the complex and carries a negative charge in the active state [82, 139, 193, 281]. The main features of anionic coordination polymerizations are [15, 219, 279, 282]:

— a head-to-tail chain,
— no branching by a chain-transfer reaction,
— no branching by copolymerization of monomers with their oligomers.

The lack of such types of branching is also explained by the well-known fact that carbanion stability decreases in the order: primary > secondary > tertiary, which is the reverse order to carbonium ion stabilities.

— the monomer unit, responsible for a particular steric configuration, corresponds to a well-defined arrangement.

The termination of anionic polymerization takes place by chain transfer to the solvent, to a monomer or to other compounds present in the reaction mixture [283]. When an inert solvent is used or in the absence of other reactive impurities, no termination occurs and a so-called "living" polymer results.

Polymerization with stereospecific catalysts yielded the greatest number of stereoregular polymers. The influence of various parameters (temperature, catalyst system, monomer concentration, solvent, various impurities, phase state) on the course of the stereospecific polymerization is generally known. The chemistry of these reactions and the dependence of stereospecific polymerization on monomer and catalyst structures are less known. Therefore, several approaches and assumptions concerning the polymerization mechanism have been made. The mechanism has not been completely elucidated; however, the most probable approach seems to be the coordination mechanism suggested by Natta, the active site involving a carbanion.

A number of experimental evidence was gathered by Natta to confirm his assumption [101, 133, 277, 284]:

— the presence of active sites within the catalyst system;
— polymerization with stabilized catalyst systems involves a constant number of active sites;
— the polymer chains contain vinylidene-type double bonds;
— the macromolecules are linear and have a regular structure;
— the activation energy does not exceed 15 kcal/mole;
— the molecular weight of the polymer is not lowered by the solvents used as reaction medium despite their ability to trap free radicals.

Although this type of mechanism and propagation through a metal–carbon bond were generally accepted, the particular metal involved in the metal–carbon bond is, however, controversial. Some authors assume that the propagation reaction occurs through the Al—C bond [69, 285] or through a bridge bond [101, 286], while others think that the Ti—C bond is involved [108, 109, 111, 114–116, 122, 235, 236, 286, 287–290]. Propagation at a surface microdefect of the insoluble transition metal compound was also suggested [113].

1. The propagation reaction at the Al—C bond

Natta was the first to suggest this type of reaction [101, 291]. An association of the monomer with the transition metal is assumed to be the first step, as the result of the high electron density of the double bond and the electron-deficient transition metal. Chain growth is attributed to a neutralization of the charge carried by the CHR group of the monomer with the CH_2 group of the polymer chain; the CH_2 group of the monomer is attached to the more electropositive metal, namely aluminium (see the reaction on p. 44).

Jones [292] proved that in the $Al(i—C_4H_9)_3 + TiCl_3$ catalyst system, titanium trichloride remained unchanged, confirming Natta's assumption that this catalyst system involves the $TiCl_3$ surface on which small amounts of AlR_3 are chemisorbed.

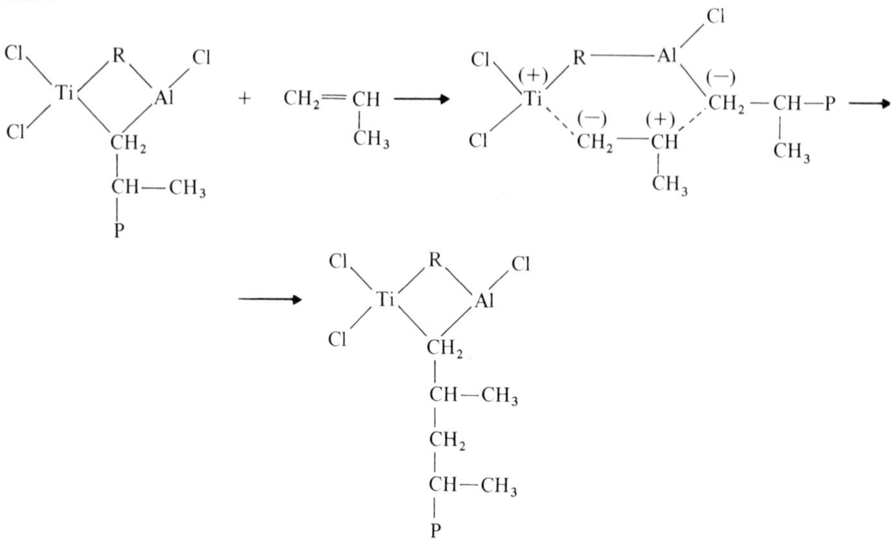

$$\mathrm{Cl_2Ti\langle R, CH_2(CH(CH_3)P)\rangle AlCl} + \mathrm{CH_2{=}CH(CH_3)} \longrightarrow \mathrm{Cl_2Ti^{(+)}\cdots R{-}AlCl{-}^{(-)}CH_2{-}CH(CH_3){-}P;\ Ti\cdots{}^{(-)}CH_2{-}{}^{(+)}CH(CH_3)\cdots Al} \longrightarrow$$

$$\longrightarrow \mathrm{Cl_2Ti\langle R, CH_2{-}CH(CH_3){-}CH_2{-}CH(CH_3){-}P\rangle AlCl}$$

Gaylord and Mark [80] proposed a mechanism slightly different from the one Natta proposed. The reaction occurs in two steps:

— monomer adsorption on a catalyst site forming an active catalyst–monomer complex,
— monomer addition to the growing polymer chain:

$$\mathrm{C + M} \overset{K_1}{\rightleftarrows} \mathrm{(CM)\ fast}$$

C = active catalyst site

M = monomer molecule

(CM) adsorbed catalyst–monomer complex

$$\mathrm{(CM)} \overset{K}{\rightarrow} \mathrm{C + polymer\ (slow,\ rate\ determining)}$$

In the case of butadiene the active complex (CM) can assume three stereochemical arrangements, yielding the three *trans*-1,4-, *cis*-1,4- and 1,2-structures.

2. *The propagation reaction at the Ti—C bond*

Contrary to Natta's hypotheses, other researchers assume the existence of a reaction mechanism involving the propagation at the Ti—C bond. In support of this mechanism, investigations were made on the stereospecific polymerization of olefins with titanium complexes. Oita and Newitt [117] proved that ethylene can polymerize at low temperatures with free radicals in the presence of $TiCl_4$ and in the absence of aluminium alkyls. Beerman and Bestian [118] succeeded in isolating methyltrichlorotitanium from the reaction of $(CH_3)_2AlCl$ with $TiCl_4$. The isolated compound was able to promote the polymerization of ethylene.

Natta and Mazzanti ([101] consider that polymerization which occurs by initiation at the Ti—C bond must have a different reaction rate compared with propagation at the Al—C bond and that experimental evidence seems to favour the growth on the Al—C bond.

Anionic coordination mechanisms have been proposed, among others, by McGowan and Ford [293] who suggested the formation of a complex by electron transfer from a trivalent alkyltitanium derivative to ethylene as the first step in the polymerization of ethylene.

According to Ludlum and co-workers [235], initiation and propagation occur through the intermediacy of an alkyltitanium(II) halide which complexes with ethylene and adds the monomer stepwise after its coordination with the active catalyst.

Dawans and Teyssié [136], Babitski and Dolgoplosk [217], and Boor [113] assumed as most probable, mechanisms assigning the main role to the reaction of the transition metal, to its valency state and to the type of ligand which stabilized this valency state. Titanium compounds react in a lower valency state IV since it provides the necessary free electrons for complex formation. The reducing metal has the role of alkylating the titanium, increasing its ability to form complexes and facilitating the redistribution of electrons.

Cossee [109] assumes a mechanism involving a monometallic active site, namely titanium.

When the reaction is repeated, the alkyl group reverts to its initial position; this is possible owing to the non-equivalence of the crystallographic positions of both chlorine atom voids and the alkyl groups on the active site. In this way the alkyl group can move back to its initial position after each monomer addition. The polymerization is in fact a series of identical steric sequences leading to the isotacticity of the product.

In a paper on the mechanism of stereospecific polymerization, Uelzman [104] thinks that metal alkyls initiate the polymerization through their metal cation and not through the carbanion. Orientation is no longer possible when only one metal atom is present, since the polymer chain can move freely changing its position after the activation of the monomer. Therefore, the necessary conditions for a constant, specific monomer orientation are fixed positions for both the cation and the anion. The pre-polarized double bond reacts with the cation, which further activates the carbanion formed at the other end and orientates it toward the complex anion. The pre-orientated methyl group is directed far from the surface, toward the chlorine atom, and is stabilized by hydrogen bonding.

Stereospecific polymerizations have also been achieved with soluble Ziegler–Natta-type catalysts.

Such types of catalysts were described in Chapter II, Section B_2. Most of the suggested models assume an anionic coordination mechanism involving propagation on the transition metal–carbon bond. Thus, Breslow and Newburg [103, 194] showed that with catalysts of the following type

$$(C_5H_5)_2Ti(Cl)\text{—}Cl\text{—}Al(Cl)(C_2H_5)_2$$

a complex between the titanium and the olefin is formed. The role of the aluminium alkyl is to alkylate the titanium, creating a positive charge and increasing the ability to form a complex. Then, the transfer of an ethyl group or of an alkyl group from the growing polymer occurs and a new Ti—C bond is formed.

An analogous mechanism was proposed by Carrick [114, 294] for the polymerization of ethylene in hydrocarbon solvents with catalysts prepared from aluminium tribromide, an organoaluminium compound and a vanadium halide:

$$(R\ \text{or}\ X)_2Al(\mu\text{-}X)_2V\text{—}R \qquad R = \text{alkyl or aryl} \qquad X = \text{halogen}$$

Polymerization occurs via the coordination of ethylene with the vacant orbitals of the vanadium, followed by a rearrangement resulting in a net addition of V—R to the double bond of the ethylene.

Polymerization with chromium acetylacetonate—AlR_3 is, according to Bawn [214], a typical anionic coordination mechanism. The fact that 1,2-polybutadiene is the prevailing structure, proves that the previous monomer coordination scheme is unlikely. According to Bawn, polymerization would proceed by the activation of a single double bond.

The most important catalysts in this class are cobalt-based yielding 99 % *cis*-1,4-polybutadiene. The mechanisms with these catalyst systems are controversial and various researchers suggested either a cationic coordination mechanism [279, 295, 296] or an anionic coordination mechanism [90, 297].

Natta's [297] and Cooper's [298] arguments in support of an anionic coordination mechanism are based on the same termination reaction with tritiated alcohol, ROH*, as Childers' [279] assumption of a cationic coordination mechanism. Polymer activity terminated with CH_3OH^* was confirmed. According to Perry [274] this is evidence for an anionic coordination mechanism.

Duck [90] also assumes an anionic coordination mechanism. In the polymerization of butadiene, a Lewis complex between alkyl aluminium chloride and the cobalt salt is first formed

$$(C_2H_5AlCl_2X)^-(CoX)^+$$

Propagation occurs at the Co—C bond and involves a negative ion which stabilizes the chain end, and a positive ion maintaining the transition metal in a low valency state, capable of forming sandwich-type bonds:

$$[RAlCl_3]^- \quad {}^+[CoX] \longrightarrow [RAlCl_3]^- \quad {}^+[CoX]$$

$$[RAlCl_3]^- + \begin{matrix} CH{=}CH \\ | \quad\quad | \\ CH_2 \quad CH_2 \end{matrix} - \quad {}^+[CoX]$$

The charge redistribution in the butadiene molecule is induced by the electrical charge on the complex. Therefore, the butadiene molecule moves in order to occupy a more energetically stable position between the Lewis complex ions.

In conclusion, stereospecific polymerization in the presence of a coordination catalyst occurs in two steps:

- the formation of a complex capable of fixing the monomer into a specific steric configuration through at least two metal–monomer bonds;
- the fixed monomer adds to the growing chain because of its particular energetic state and under the influence of a physical or chemical agent.

Chapter IV

Kinetics of stereospecific polymerization

ADDITION polymerization is a chain reaction in which a few active sites induce a sequence of similar reactions.

From a kinetic point of view, the formation of a macromolecule involves the following steps [299]:

— the initiation reaction resulting in the formation of the active species;
— the chain propagation reaction. In kinetic studies the assumption is made that the rate of formation and that of consumption of the chain carrier are almost equal. In this reaction step a quasi steady state is reached and the polymerization rate is considered constant. This assumption is true only if the propagation rate is higher than the initiation rate.

The average size of the polymer molecule is determined by the rate of the growth-reaction and the rates of the competitive termination reactions, i.e. chain transfer or isomerization, and is given by the following ratio:

$$\frac{R_p}{R_b} = R_p/(R_t + R_{tr}),$$

where R_p is the propagation rate, R_b the overall rate of all chain termination reactions, R_t the rate of simple termination and R_{tr} the rate of chain transfer.

— the chain transfer to monomer; this is a strongly temperature-dependent reaction especially in cationic polymerizations.
— transfer reactions from the growing chain to other substances present in the system, namely solvent, initiator or special additives introduced as transfer agents; these are the main transfer reactions encountered in stereospecific polymerizations;
— chain termination reactions occurring by chain recombination in free radical polymerizations. This type of termination is not possible in ionic polymerizations owing to electrostatic repulsion between identical charges. In stereospecific polymerizations, "living" polymers, without termination step, described by Szwarc [246, 247], are usual.

Fontana [300] emphasized the great number and complexity of the reactions occurring in ionic polymerizations both in the propagation step and in the initiation, transfer or termination steps, because:

— the active species is usually a polar complex, the counter-ion playing an important role in determining the reaction type;
— a strong interaction of the active species with the solvent arises as the result of polar effects;
— ion dissociation or agglomeration may induce certain secondary effects;
— in certain systems, a specific reversible association of the active species with the monomer occurs.

The main factors affecting the reaction kinetics are the dependence on reagent concentration, temperature and certain particular properties of the solvent such as the dielectric constant and the viscosity.

The kinetic equations show the differences between free radical and ionic polymerizations, i.e. the polymerization rate and the low activation energy, sometimes negative, of ionic polymerization as compared with radical polymerization [301]. Thus, the stereospecific polymerizations of some olefins and dienes with Ziegler-type catalyst systems have activation energies ranging from 1 to 15 kcal/mole; only seldom do they reach higher values up to 22 kcal/mole [302–305].

Kinetic studies have been made of ethylene [236, 293, 306–308], propylene [27, 137, 139, 309–312], 1-hexene [313], styrene [302, 314–316], isoprene [317], and butadiene [172, 225, 318–320]. Similar values for the activation energies were also found in polymerizations with lithium catalysts [305, 306, 321].

Kinetic studies were made on certain stereospecific polymerization systems with the object of confirming one of the possible reaction mechanisms.

Taking into account the influence of some impurities, even in amounts of parts per million, on the polymerization process and on the structural features of the resulting polymers, it is obvious that it is very difficult, and sometimes even impossible to obtain, accurate qualitative and quantitative kinetic data and to isolate and characterize certain intermediates. As shown above, these difficulties are due largely to the lack of precise data concerning the structure of the products formed in the reaction of $TiCl_4$ with AlR_3 under certain given conditions. Several types of compounds are probably formed and only some of them (or possibly only one) are responsible for the particular catalytic activity. According to Sigwalt [239] only 0.1 % of the $TiCl_3$ present in the system gives rise to active sites. Moreover, the catalyst composition can change during the polymerization reaction.

Most kinetic studies have been made with catalyst systems which are assumed to act via an anionic coordination mechanism; in numerous papers,

Natta reported experimental kinetic evidence for such a mechanism [101, 133, 277].

The complexity of the kinetic studies of ionic or ionic coordination polymerization reactions arises from the difficulty of clearly separating the various polymerization steps. Usually, the overall polymerization rate has been determined, but even here, various factors, such as the degree of conversion, interfere and produce erroneous results.

Initiation of polymerization on a catalyst surface complicates the kinetic studies. The growing macromolecules attached to the active sites of the catalyst surface also interfere in kinetics measurements.

The most important factors affecting the kinetics of polymerization are the following:

1. The activity of the catalytic sites

Both the constant monomer concentration and the constant number of active sites during the propagation step are highly important for accurate kinetic measurements. Polyethylene, with a very large distribution of molecular weights, is obtained with Ziegler–Natta-type catalysts [321–325]. This polydispersity may be ascribed to the presence within the Ziegler–Natta-type catalyst system of catalytic sites with different activities, each of them having its own rate constant (k_p) or transfer constant (k_{tr}) [76, 326]. Therefore, the resulting polymer is a mixture of various types of macromolecules, each macromolecular chain resulting from a certain active site and having a specific distribution of molecular weights. Hence, the overall distribution function is the sum of several separate distributions.

The chemical deactivation of the catalytic sites by poisoning with certain secondary products formed during the polymerization reaction produces changes in the polymerization rates at conversions higher than 10 or 20% [327].

2. The ionic nature of the catalyst system

Ionic catalyst systems influence both the polymerization rate and the product structure by steric and ionic factors involved in the initiation step.

The simple anionic initiation results in the formation of a 1,2-structure in the case of butadiene and of 1,2- and 3,4-structures in the case of isoprene and of various amounts of *cis*- and *trans*-1,4 structures [328]; however, lithium alkyls, which in aliphatic hydrocarbon solvents form ionic coordination compounds, produce *cis*-1,4-polyisoprene. At higher concentrations and in aromatic solvents, lithium alkyls also favour the formation of products with a

3,4-structure [328]. Under such conditions the anionic coordination character of the system is reduced.

3. *The influence of the Al/Ti ratio and of the concentration of the catalyst*

In the polymerization of butadiene with the $Al(i\text{-}C_4H_9)_3 + TiCl_4$ catalyst system in heptane at 25°C, the monomer consumption rate was dependent on the Al/Ti ratio and on the catalyst concentration. At Al/Ti molar ratios ranging from 0.5 to 1.25, irrespective of the concentration of both catalyst components, a polymer with 96–99 % *trans*-1,4-structure was obtained. The highest polymerization rate was reached at an Al/Ti molar ratio of 2.7. However, at this Al/Ti molar ratio, the polymer contained 75–80 % of insoluble gel and 20–25 % of a benzene-soluble fraction. The soluble fraction contained about 65 % *trans*-1,4 product and 30 % *cis*-1,4. At Al/Ti molar ratios ranging from 1 to 1.6, the monomer consumption rate was first order with respect to butadiene concentration. At Al/Ti molar ratios higher than 2 the monomer consumption rate became second order with respect to butadiene concentration. At a constant Al/Ti ratio, the polymerization rate was proportional to the catalyst concentration [329].

Isoprene yields *cis*-1,4 polymers at a Al/Ti molar ratio equal to one. The amount of *trans*-1,4 product increases at lower Al/Ti molar ratios [317, 330, 331].

4. *The influence of the solvent*

The nature of the solvent, especially its dielectric constant, has a marked influence on the rate of stereospecific polymerization and, sometimes, also on the structure of the product.

A 96.5 % conversion was reported for the polymerization of butadiene in benzene solution with soluble cobalt catalysts. The degree of conversion gradually decreases with an increase in the dielectric constant of the solvent, the conversion being only 26.1 % in *metha*-dichlorobenzene with a dielectric constant of 10.2. Such solvents have, however, quite a reduced influence on the microstructure of the polybutadiene [295].

By working with the same cobalt catalyst system, if benzene is partly replaced by aliphatic or cycloaliphatic hydrocarbons [332, 333], the molecular weight of the *cis*-1,4-polybutadiene decreases while the polymerization rate increases, since in saturated hydrocarbons the complex formed between Lewis bases and ethylaluminium sesquichloride is largely dissociated and is more active as a chain termination agent than in benzene solution [213].

Adams *et al.* [331] obtained *cis*-1,4-polyisoprene with the $AlR_3 + TiCl_4$ catalyst system in toluene/petroleum ether. In 1-chloropentane, an explosive cationic polymerization occurs and a white solid resin is obtained. In *ortho*-dichlorobenzene, a slightly acid solvent, with a dielectric constant of 7.5, polyisoprene is formed at an Al/Ti molar ratio greater than unity.

With lithium catalysts, the influence of the solvent on the polydiene microstructure is striking [334]. Isoprene polymerization with lithium alkyls or other organoalkali compounds in polar solvents results in the formation of a polymer with 1,2-, 3,4- and *trans*-1,4 structure [335]. In benzene, diphenyl ether and anisole, a mixture of *cis*-1,4- (93–64 %) and 3,4 (7–36 %) polymer is produced. In ethyl ether and tetrahydrofuran, no *cis*-1,4-polyisoprene is formed. Only the addition of electron donor compounds reduces or even completely destroys the catalyst stereospecificity [177, 336].

Activation energies are also dependent on the solvent type; diene (isoprene, butadiene) polymerization in *n*-hexane has an activation energy of 27 kcal/mole, while in tetrahydrofuran the energy is only 6.7 kcal/mole [231].

5. *Diffusion phenomena*

An important factor influencing the ratio of polymerization is the physical state of the catalyst and the time the polymer remains attached to the active catalyst sites. The actual causes for the decrease in rate at conversions higher than a given limit are not known precisely. It may be produced by an agglomeration of polymer particles preventing the monomer diffusing to the catalytic sites by blocking the catalyst [220, 302, 337–340]. In order to maintain a constant rate of polymerization, the entry of the monomer into the catalyst system should not be prevented by the polymer. This is only possible at low conversions. The above remarks seem to be corroborated by the molecular weight distribution. At very low conversions, the molecular weight distribution is relatively narrow, but it rapidly widens with an increase of the conversion. Diffusion phenomena may interfere with normal polymerization, favouring the formation of smaller molecules as a result of their rapid removal from the active surface [76].

On the other hand, Fukui [341] completely discards the theory of diffusion phenomenon. Grieveson [301], who followed the polymerization course microscopically, assumes that the macromolecule carries away the catalyst during the growing step. It means that cleavage of the initial catalyst aggregates occurs during the polymerization reaction. This deaggregation means that the catalyst particles are not crystals of $TiCl_3$ as such but rather loose aggregates of mycrocrystalline material. From this microscopic investigation, it is concluded that the coating of the catalyst surface with a polymer layer does not take place.

Kinetic studies have been performed with the following systems:

Monomer	Catalyst system	References
Ethylene	$Al(C_2H_5)_3 + TiCl_4$	[235, 342]
	$Al(C_2H_5)_2Cl + TiCl_3$	[301, 326]
	$Zn(C_4H_9)_2 + TiCl_4$	[249, 225]
	$LiR + TiCl_4$	[80]
Propylene	$Al(C_2H_5)_3 + TiCl_3$	[137, 139, 343–346]
	$AlR_3 + TiCl_3$	[309, 347]
	$Al(C_2H_5)_2Cl + TiCl_3$	[347]
Butadiene	$Al(i\text{-}C_4H_9)_3 + TiCl_4$	[348]
	$AlI_3 + TiCl_4 + AlHCl_2(C_2H_5)_2O$	[318]
	$Cr(acac)_3 + Al(C_2H_5)_3$	[214]
	$Co(acac)_2 + AlR_2Cl$	[214, 319]
	$Co(acac)_2 + Al_2R_3Cl_3$	[213]
	$Ti(OC_4H_9)_4 + Al(C_2H_5)_3$	[225]
	$CoCl_2Py + Al(C_2H_5)_2Cl$	[319, 320]
	$CoCl_2Py + Al(C_2H_5)_2Br$	[319]
	$CoCl_2Py_2 + Al(i\text{-}C_4H_9)_2Cl$	[349]
	$Al(i\text{-}C_4H_9)_3 + TiI_4$	[312]
	BuLi	[192, 350, 351]
	C_2H_5Li	[305]
Isoprene	$Al(i\text{-}C_4H_9)_3 + TiCl_4$	[80, 317, 352]
	$Al(C_2H_5)_3 + VCl_3$	[127]
	$TiCl_4 + LiR$	[80]
	BuLi	[192, 195, 353–355]
	Lithium metal	[303]
	Polyisoprenyl lithium	[356]
Styrene	BuLi	[192, 351, 355, 357, 358]
2,3-Dimethylbutadiene	Lithium metal or lithium alkyl	[315]
cis-1,3-Pentadiene	Lithium metal	[359]

acac = acetylacetonate; Py = pyridine.

The most useful experimental techniques are dilatometric methods, evaluating the rate from the consumption of monomer, and gravimetric methods, measuring the polymerization rate by the amount of polymer obtained.

Ludlum [235], as well as Fellchenfold [342], suggested for ethylene polymerization with the $AlR_3 + TiCl_4$ catalyst system, the following kinetic equation:

$$R_p = -\frac{d[E]}{dt} = k_p[E][C]$$

where R_p is the rate of propagation, k_p is the propagation rate constant, [E] the ethylene concentration, and [C] the concentration of the active sites.

According to McGowan [293] and Gilchrist [249], the polymerization is second order with respect to the monomer, first order with respect to $TiCl_4$ and zero order with respect to metal alkyl, in the presence of the $TiCl_4 + Zn(n\text{-}C_4H_9)_2$ catalyst system.

A first-order polymerization with respect to the concentration of the alkylated titanium complex was also estimated by Breslow [360]. The reaction is first order with respect to the monomer, in the polymerization of 1-butene [361], styrene [362], isoprene and butadiene [329], with Ziegler–Natta-type catalysts.

In his studies on the kinetics of ethylene polymerization with the $Al(C_2H_5)_2Cl + TiCl_3$ catalyst (stable for several hours in certain given conditions without chemical deactivation or coating the active sites with polymer), Grieveson [301, 326] also found a first order reaction with respect to both monomer and $TiCl_3$ at 30 and 40°C. At 60°C, a second-order reaction with respect to the monomer concentration was estimated. A zero-order reaction with respect to aluminium alkyl concentration was found. The reaction rate increased with temperature up to 50°C, but then decreased with a further increase of temperature.

Schindler [363] found for the propagation rate of the ethylene $-Al(i\text{-}C_4H_9)_2H + TiCl_4$ system the following equation

$$R_p = \frac{k_p[E]^2}{1 - k'_p[E]}$$

where k_p and k'_p were dependent on the diluent gas. The adsorption of ethylene on the active catalytic sites was the rate-determining step.

Similar relationships were found by Fukui [341] using the $AlEt_3 + TiCl_3$ catalyst system in *n*-heptane, but the polymerization rate increases with increase in concentration of $Al(C_2H_5)_3$.

Kinetic studies on the polymerization of propylene with Ziegler–Natta-type catalysts suggested a reaction mechanism based on certain particular features. The initiation reaction provides the coordination of the monomer with the transition metal of the active catalyst site [364]; propagation takes place on this intermediate. The reaction sequence shows no kinetic termination [365] and the chain size is determined by the rates of the transfer reaction to either the monomer or the organometallic compound [93, 96, 97]. The same active site may initiate several macromolecular chains.

The polymerization rate of propylene (which is equal to the propagation rate) is expressed by the following equation:

$$R_p = k_p[TiCl_3][C_3H_6][Al]^o$$

The reaction is first order with respect to both monomer and titanium trichloride and under certain conditions it is independent of the concentration of the aluminium compound [27, 33, 139, 366].

The constant propagation rate indicates the absence of an important chain-termination process. The last reaction is not detected kinetically since it is counter-balanced by the formation of a new active site [338].

Ternary catalyst systems, like $Al(C_2H_5)_3 + TiCl_3 + TiCl_4$, give no reproducible results. The polymerization rate is not constant and it suddenly drops if the catalyst is not preformed [340].

Few kinetic studies were made on diene polymerization with Ziegler–Natta-type catalysts.

Saltman [317] reported that the reaction rate is first order with respect to the isoprene concentration at a constant Al/Ti molar ratio, using the $TiCl_4 + Al(i\text{-}C_4H_9)_3$ catalyst system. When replacing $TiCl_4$ by TiI_4, the reaction order with respect to monomer remains unchanged while the microstructure is markedly modified [319].

A first-order reaction with respect to butadiene concentration was also calculated when polymerization was carried out with soluble modified Ziegler–Natta-type catalysts. The reaction is first order (seldom 1/2) with respect to the cobalt compound [349], while the reaction order is 1/2 with respect to the aluminium compound [319, 320].

The use of certain catalyst systems containing no Al—C bonds, such as $TiCl_4 + AlI_3 + AlHCl_2 \cdot (C_2H_5)_2O$ [318], for the homogeneous polymerization of butadiene in toluene, gives a first-order reaction with respect to the concentration of both the monomer and the catalyst.

In the polymerization of butadiene with chromium triacetylacetonate and $Al(C_2H_5)_3$ in homogeneous medium, the initiation is instantaneous [172]. The initial rate of polymerization of isoprene with $VCl_3 + Al(C_2H_5)_3$ [127] is slow and is followed by an acceleration up to a steady rate. This acceleration is assigned to a dissociation of the crystalline VCl_3 aggregates induced by the growing chains. Only $\frac{1}{160}$ of the vanadium atoms of the unchanged catalyst are involved in polymerization.

Strong electron donor compounds show an inhibiting effect which is assumed to be determined by an interaction between the monomer and the donor taking place on the surface of the active site [127]. Weaker donors (aliphatic ethers or thioethers) do not reduce the rate and sometimes the rate is even increased (probably by a stabilization of certain specific active sites).

Most kinetic studies have been undertaken with styrene, isoprene and butadiene polymerized with lithium metal or lithium alkyls, especially butyllithium. In the case of styrene, the initiation involves the addition of *n*-butyllithium to the double bond of the monomer [367]. Each initiator molecule yields a macromolecular chain [368]. At a high monomer/catalyst ratio, initiation is complete before the whole of the monomer is consumed.

In polymerizations carried out at a low monomer/catalyst ratio the rate of styrene consumption depends on both the initiation and the propagation rates; a second-order reaction with respect to the monomer concentration and a zero order in catalyst have been recorded [179]. Welch [369], who measured the propagation rate during the last stages of the reaction, found a first order reaction with respect to monomer. Chain initiation, estimated from the initial rate of formation of the ionic intermediate, is first order with respect to monomer and 1/6-order with respect to *n*-butyllithium. The propagation is also first order with respect to monomer, but only 1/2-order with respect to active sites [368].

Isoprene polymerization with lithium alkyls in benzene, *n*-hexane or in bulk [370] shows a slow initiation rate. The chain propagation is first order with respect to monomer. Morton [304] reported that the propagation is 1/2-order with respect to active sites, in close agreement with Bywater [368]. This result is explained by the association of the polymeric ion pair, only the dissociated species being effective in the polymerization process.

In tetrahydrofuran, the propagation step involves the reaction of a simple ion pair with the monomer. The overall rate, which is equal to the propagation rate for rapid initiation steps, is first order with respect to monomer and to active site concentrations; the activation energy is 7 kcal/mole.

Similar kinetics were evaluated for the polymerization of butadiene with lithium metal and lithium derivatives.

In a comparative study on the polymerization of styrene with *n*-, *sec*-, *tert*- and *i*-butyllithium, Hsieh found the equation: $R_i = k_i[\mathrm{BuLi}][\mathrm{M}]$ for the initiation rate. The rate constants change with the monomer, the solvent and the organolithium compound. The conversions are a function of the active site concentration. The higher the lithium compound concentration, the more associated the organolithium compound and the less active the catalyst site.

A controversial question, not yet solved, with regard to lithium-initiated polymerization, is the reaction order of the polymerization with respect to the lithium or alkyllithium concentration. Although a first-order reaction with respect to monomer is generally accepted for the initiation and propagation, for the reaction order with respect to the metal alkyl concentration various values have been found, namely $\frac{1}{2}$ [37], $\frac{1}{3}$ [357], $\frac{1}{5}$ and $\frac{1}{6}$ [335, 355, 372].

Sinn [335] devised for the overall polymerization reaction of ethylene or isoprene with metal alkyls the following equation

$$V_{\mathrm{overall}} = k[M]\sqrt[n]{C_o}$$

where C_o is the total concentration of the organometallic compound as determined by titrimetric methods, n is the number of moles of metal alkyls in the associated aggregate, namely two for aluminium alkyls and six for lithium alkyls, and $[M]$ is the ethylene or isoprene concentration. Sinn assumed that only the monomeric species of the organometallic compounds were active in initiation.

The investigation of the propagation rate is also highly important. It was shown that in most cases, the real initiators in polymerization with lithium alkyls are not the simple alkyl derivatives but lithium polyalkenyls formed during the first stage of the polymerization and having molecular weights ranging from 5000 to 10,000.

It was proved that such alkenyllithium species are associated into pairs. This is true for styrene in benzene solution and for isoprene and butadiene in hydrocarbon solvents [371], but in tetrahydrofuran solution no association occurs.

The role of lithium alkenyls in chain propagation was precisely established by separate investigation of the propagation and initiation rates. The pre-initiation or seeding procedure was used in this work, consisting in the introduction of previously prepared lithium alkenyls into the polymerization system, initiating instantaneously the propagation reaction [183, 354, 358, 371–376].

Initiation with butyllithium has a slow initiation rate and therefore initiation and propagation occur simultaneously. On account of this observation, Hsieh avoids using n-butyllithium for the preparation of lithium alkenyls [192].

The use of lithium metal to prepare initiators by reaction with the monomer has this advantage: after the formation of the lithium alkenyls, excess lithium metal can be removed from the mixture by simple filtration and, in this way, the initiation and the propagation steps can be studied separately [377].

Bywater [351] ascribed the fractional reaction order with respect to lithium alkyl concentration to the association of the lithium polyalkenyls which are the true initiators and not to the association of the lithium alkyl. A dimer form is assigned for polystyrenyllithium, a tetramer for polyisoprenyllithium and a hexamer for polybutadienyllithium. The degree of association is probably dependent on the geometry of the substituent in the carbanion.

In agreement with Bywater's assumption, Brown [182] does not accept the monomeric lithium alkyl as the active kinetic species and assigns this role to the new organolithium compound with results from the reaction between RLi and olefin.

Waack [184] and others [179, 186–189, 378] think that for olefin polymerization in hydrocarbon solvents a 1/6 reaction order with respect to n-butyllithium may be attributed to the reactive form of monomeric n-butyllithium in equilibrium with a hexameric form, relatively inactive and in excess.

Margerison's measurements [181] prove the preponderant existence in solution of the hexameric butyllithium species.

The above experimental data emphasize the diversity of approaches in elucidating stereospecific polymerization mechanisms. In attempting to compare the mechanisms suggested for identical systems, e.g. $TiCl_4 + AlR_3$, for which Natta suggested an anionic coordination mechanism, Uelzmann proposed a cationic initiation and an anionic propagation mechanism, while Friedlander favoured a radical ionic mechanism, a decision has to be made which arguments are decisive. The same difficulties of deciding between the ionic, radical or radical-ionic character of the intermediate species are also found in isoprene polymerization in both the adsorbed state and in the liquid phase. With the $TiCl_4 + Al(i\text{-}C_4H_9)_3$ catalyst system, the formation of an intermediate ionic species seems more probable. Kinetic evidence, gathered by Natta for this catalyst system, strongly supports an anionic coordination mechanism. However, further investigations are needed to confirm or invalidate propagation on the transition metal–alkyl bond.

The above various approaches to the polymerization mechanism and kinetics call for a careful investigation of all the factors involved in stereospecific polymerization (catalyst species, reagent type and purity, working conditions, etc.) capable of influencing to some extent the reaction mechanisms and implicitly the microstructure and other properties of the resultant polymers.

EXPERIMENTAL PART

Chapter I

Purification of the reagents used in the stereospecific polymerization of isoprene

STUDIES on the purification of the reagents used in the stereospecific polymerization of isoprene are highly important as the purity of the monomer, solvent and inert gas is essential for the preparation of a highly stereoregular polymer.

For polymerization studies, two isoprene fractions with concentrations of over 99 % and *ca.* 25 % obtained by the dehydrogenation of the same raw material, isopentane, were used.

As a result of the synthesis procedure, the monomer contains certain impurities such as: cyclopentadiene, carbonyl compounds, acetylenic hydrocarbons, oxygen and water which are detrimental to the polymerization process.

The solvents used as a reaction medium in polymerizations of isoprene of purity over 99 % were saturated aliphatic hydrocarbons, *n*-heptane and a hexane fraction, containing oxygen and water as undesirable impurities.

Two heterogeneous Ziegler–Natta-type catalyst systems were used with triethylaluminium (TEA) or triisobutylaluminium (TIBA) as the aluminium alkyl component and titanium tetrachloride, as the transition metal derivative. The aluminium alkyls had the required grade of purity for use in polymerization reactions. Titanium tetrachloride, synthesized by the chlorination of titanium dioxide in the presence of carbon, had a purity of 99.9 %; it was distilled before use for the preparation of the catalyst complex.

The purity of nitrogen, used as the inert gas, was over 99.5 %, but it still contained traces of oxygen and water as impurities.

The content of impurities of the reagents, determined by chemical methods, was within the ppm range, but still well above the permitted levels for stereospecific polymerization.

It was therefore necessary to find the most efficient methods to remove each impurity.

A. Purification of isoprene

The two isoprene fractions with purities of over 99 % and *ca.* 25 %, obtained by dehydrogenation of isopentane, had the following physical constants:

	>99 % isoprene	~25 % isoprene
n_D^{20}	1.4220	1.3721
d_4^{20}, kg/m^3	682	660
b.p. at 760 mm Hg	34.1°C	28–34.1°C

The chemical composition of both isoprene fractions was determined by gas chromatography. The results are shown in Table 1 and in the chromatograms given in Figures 1 and 2.

Table 1

GLC analysis of over 99 % and *ca.* 25 % isoprene fractions

Compound	over 99 % isoprene (%)	*ca.* 25 % isoprene (%)
Propane	0.001	—
Propylene	0.004	—
i-Butane	—	0.007
n-Butane	0.048	0.276
1-Butene + *i*-butene	0.035	0.107
trans-2-Butene	0.024	0.087
cis-2-Butene	0.009	0.225
Isopentane	0.131	60.231
3-Methyl-1-butene	0.024	—
n-Pentane	0.108	2.175
1-Pentene	0.006	0.294
2-Methyl-1-butene	0.142	3.730
trans-2-Pentene	0.023	0.178
cis-2-Pentene	0.007	0.063
2-Methyl-2-butene	0.043	6.372
Isoprene	99.396	26.036
Piperylene	—	0.219

The chromatographic data indicate that the isoprene sample of over 99 % purity contained about 1 % of saturated hydrocarbons and olefins in approximately equal parts. The *ca.* 25 % isoprene fraction contained 63 % of isopentane, 11 % of olefins, mainly methylbutenes, and 0.2 % of other dienes.

A comparison between the two chromatograms led to the conclusion that the two monomer fractions were different both in respect of the isoprene

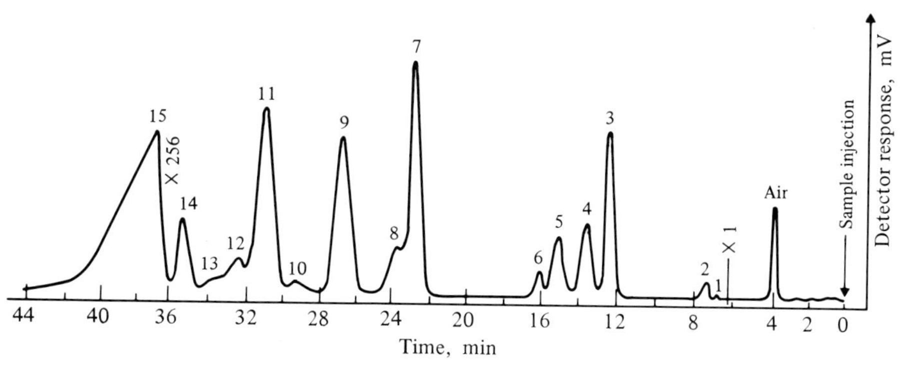

FIGURE 1. Chromatogram of isoprene fraction of over 99 % purity. 1—propane; 2—propylene; 3—*n*-butane; 4—1-butene + *iso*-butene; 5—*trans*-2-butene; 6—*cis*-2-butene; 7—*iso*-pentane; 8—3-methyl-1-butene; 9—*n*-pentane; 10—1-pentene; 11—2-methyl-1-butene; 12—*trans*-2-pentene; 13—*cis*-2-pentene; 14—2-methyl-2-butene; 15—isoprene.

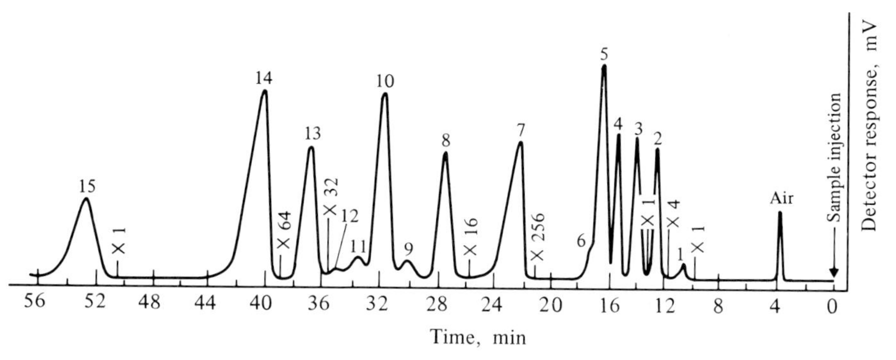

FIGURE 2. Chromatogram of isoprene fraction of purity *ca.* 25 %. 1—*n*-butane; 2—*iso*-butane; 3—1-butene-*iso*-butene; 4—*trans*-2-butene; 5—*cis*-2-butene; 6—1,3-butadiene; 7—isopentane; 8—*n*-pentane; 9—1-pentene; 10—2-methyl-1-butene; 11—*trans*-2-pentene; 12—*cis*-2-pentene; 13—2-methyl-2-butene; 14—isoprene; 15—piperylene.

concentration and of the content of saturated hydrocarbons and olefins. The polymerization of the isoprene of over 99 % purity was carried out in solution by using a saturated hydrocarbon as solvent at monomer/solvent ratios ranging from 1:3 to 1:5 (v/v). For the polymerization of the *ca.* 25 % isoprene fraction, the addition of another solvent as a reaction medium was unnecessary since the isopentane present in the monomer fraction acted as a solvent.

The olefins present in the *ca.* 25 % isoprene fraction had no influence on the course of the polymerization reaction, at the Al/Ti molar ratio employed. In the reaction conditions used, isoprene was selectively polymerized and other olefins remained unchanged in the final mixture together with the solvent.

Specific chemical methods were applied to analyse both isoprene fractions (over 99 and *ca.* 25 %), to determine some of the components, in ppm amounts, which were also contaminants for the polymerization process. The results obtained by chemical analysis are given in Table 2. It is evident that both isoprene fractions contained the same type of contaminants in about equal amounts; certain impurities, such as cyclopentadiene and carbonyl compounds were present in amounts ranging within rather large limits. The concentrations of other impurities like oxygen, water and acetylenic hydrocarbons were relatively constant.

Table 2

Chemical analysis of over 99 % and *ca.* 25 % isoprene fractions

Compound	over 99 % isoprene (ppm)	*ca.* 25 % isoprene (ppm)
Cyclopentadiene	50–500	40–500
Oxygen	25–30	20–30
Water	25–40	30–40
Carbonyl compounds	40–300	50–300
Acetylenic hydrocarbons	20–30	20–30

The permissible amounts of impurities in the preparation of a polymer with a high content of *cis*-1,4 units were determined experimentally. The results obtained led to the conclusion that the deleterious impurities present in isoprene should not exceed 10 ppm for each component.

The chemical analysis of both isoprene fractions indicated that the original impurity content was much higher than the limits permissible in polymerization processes. An investigation was therefore necessary in order to reduce the impurity content to below the established limits. Physical and chemical methods were used to remove the impurities. Cyclopentadiene was removed by a chemical method, while physical methods were used for the removal of oxygen, inhibitor, water, carbonyl compounds and acetylenic hydrocarbons.

1. Removal of cyclopentadiene from isoprene

Cyclopentadiene is one of the most effective inhibitors of the stereospecific polymerization of isoprene, if present in amounts higher than 10 ppm [379].

The Diels–Alder reaction (diene synthesis) of both isoprene and cyclopentadiene with maleic anhydride is well known. However, cyclopentadiene reacts at room temperature or even below, while isoprene reacts only at the higher temperature of about 100°C [380, 381]. The removal of cyclopentadiene from isoprene is based on this reactivity difference and was undertaken by treatment with maleic anhydride at a low temperature. A solid Diels–Alder adduct containing a bridged six-membered ring, easily separated from isoprene, is formed in the reaction.

The reaction of cyclopentadiene with maleic anhydride is as follows:

```
      CH                                          CH
    //  |                                       /  |  \
  CH    |              CH——CO               CH     |    CH——CO
  |     CH2  +         ||       \O   ——>    ||    CH2   |       \O
  CH    |              CH——CO /             CH     |    CH——CO /
    \\  |                                       \  |  /
      CH                                          CH
```

Isoprene containing 500 ppm cyclopentadiene was mixed with solid maleic anhydride at 5°C and the resulting adduct was easily separated by decantation and distillation. The amount of maleic anhydride and the contact time necessary to reduce the level of cyclopentadiene in the isoprene to 10 ppm were experimentally determined.

Various amounts of maleic anhydride, i.e. 0.5, 1 and 2% (w/w) based on the isoprene, and various reaction times, i.e. 3, 6 and 12 hours have been used to find out the conditions for a virtually complete removal of cyclopentadiene. The experimental results are given in Table 3.

The experimental data showed that the content of cyclopentadiene in both isoprene fractions can be reduced from 500 to 10 ppm by reaction with 0.5% maleic anhydride for 3 hours at 5°C.

The cyclopentadiene content was determined by photocolorimetric analysis of the phenylfulvene resulting from the reaction of cyclopentadiene with benzaldehyde in the presence of alkalies [382].

Table 3

Effect of the amount of maleic anhydride and of the contact time on cyclopentadiene removal (initial content of cyclopentadiene in isoprene 0.500 ppm)

Maleic anhydride g/100 g isoprene	Contact time (hr)	Cyclopentadiene, ppm	
		Isoprene	
		>99%	~25%
2	12	8	7
2	12	7	9
1	12	9	8
1	12	7	7
0.5	12	10	9
0.5	12	8	10
0.5	6	8	7
0.5	6	9	9
0.5	3	7	8
0.5	3	8	7

In conclusion, the above method for removal of cyclopentadiene from isoprene by means of maleic anhydride is simple and effective.

2. *Removal of oxygen from isoprene*

The removal of oxygen (deaeration) from isoprene was achieved by distillation with a 5-TP column under purified nitrogen. The monomer was collected at 34.1°C/760 mmHg, when working with the isoprene fraction of over 99% purity and in the 28–34.1°C/760 mmHg interval for the *ca.* 25% fraction. As well as the oxygen the inhibitor (*p-tert*-butylcatechol or hydroquinone) was also removed, remaining in the distillation flask.

Both isoprene fractions, originally containing about 30 ppm oxygen, contained only 5 ppm oxygen after distillation. The removal of oxygen from isoprene was verified by Hersch's electrochemical method [383] which consists in the measurement of the current produced in a Ag/KOH (diaphragm)/Pb cell by the presence of oxygen. The current intensity is proportional to the oxygen concentration.

The removal of the inhibitor from the distilled isoprene was controlled by a photocolorimetric method. The method is based on the colour reaction of traces of inhibitor with phosphomolybdic acid [384]. The analytical data showed the complete absence of the inhibitor in the distilled isoprene.

3. *Removal of moisture from isoprene*

In order to remove traces of water from isoprene, the use of specific sorbents was investigated. Silica gel and 4A and 5A molecular sieves were used.

Both distilled isoprene fractions contained about 40 ppm water, as determined by the Karl Fischer method [385].

(*a*) *Removal of moisture with silica gel.* The silica gel used to remove the moisture had the following characteristics:

—particle size, ϕ mm	2–5
—bulk density, kg/m^3	800
—water adsorption capacity, % (w/w)	31

Silica gel should be activated before use to remove the adsorbed water completely in order to make use of its maximum adsorption capacity. Silica gel activation was carried out at 180–200°C, for 2–3 hours. Higher temperatures were avoided since the crystal lattice of the sorbent is changed and a complete and irreversible loss of activity occurs [386].

Moisture removal from both isoprene fractions was investigated by using a contact time of 20 hours and an amount of activated silica gel of 1.5% and 3% based on the isoprene. In the experiments, isoprene samples were used with refractive indices of $n_D^{20} = 1.4220$ (over 99 % pure) and 1.3721 (*ca.* 25 % pure), respectively.

The results are listed in Table 4.

Table 4

Drying of >99 % and ~25 % isoprene on activated silica gel. Contact time: 20 hours

g SiO_2/100 g isoprene	Moisture, ppm	
	Before drying	After drying
	>99 % isoprene	
1.5	40	9
1.5	40	10
3	40	9
3	40	9
	~25 % isoprene	
1.5	38	9
1.5	38	8
3	38	10
3	38	9

From the data shown, we concluded that after a contact time of 20 hours with 1.5 % activated silica gel, the moisture content of both isoprene fractions was reduced to about 10 ppm.

The contact time required for drying both isoprene fractions with 1.5 % of activated silica gel was also studied.

The data shown in Table 5 indicate that drying down to 10 ppm can be achieved in 1 hour with 1.5 % of activated silica gel. A longer contact time did not improve the drying.

Table 5

Effect of contact time on drying of over 99 % and *ca.* 25 % isoprene fractions on 1.5 % activated silica gel

Contact time (hr)	Moisture, ppm			
	over 99 % isoprene		*ca.* 25 % isoprene	
	Before drying	After drying	Before drying	After drying
15	40	9	38	10
10	40	8	38	9
5	40	9	38	9
3	40	10	38	8
1	40	8	38	9

(*b*) *Removal of moisture with molecular sieves.* The removal of moisture from isoprene with 4A and 5A molecular sieves was also investigated.

Molecular sieves are crystalline alkali or alkaline earth aluminosilicates with the following general chemical formula [387, 388]:

$$Me_{2/n}O \cdot Al_2O_3 \cdot xSiO_2 \cdot yH_2O$$

where n is the valency of the metal (usually Na, K or Ca) and x takes values from 2 to 10 depending on the nature of the compound.

Molecular sieves consist of three-dimensional lattices formed by SiO_4 and AlO_4 tetrahedra, the negative charges of which are balanced by alkali metal or alkaline earth cations.

The crystal lattice contains cavities linked by channels of uniform sizes which permit the cavities to be filled only by those molecules whose critical cross-sections are lower than the cross-sections of the channels.

The molecular sieves have certain specific features, determined by their regular crystal structure and hetero-ionic character, which make them preferable for the purification of isoprene from moisture, carbonyl compounds and acetylenic hydrocarbons:

- they adsorb molecules with a critical cross-section smaller than the channel cross-section of the molecular sieves, which enables a good separation of liquid and gaseous mixtures to be made;
- they have high affinity and selectivity for polar compounds;
- they also have a high adsorption capacity for impurities at ppm concentrations, which ensures a high purity. Chemical inertness, mechanical strength, as well as regeneration without loss of activity, are other major advantages of molecular sieves.

First of all, the drying efficiency of the molecular sieves was experimentally tested with both isoprene fractions.

The molecular sieves were activated at 300–350°C for 6–8 hours under a stream of purified nitrogen in order to remove the desorbed water vapours.

The characteristics of 4A and 5A molecular sieves are given in Table 6.

Table 6
Properties of molecular sieves

Property	Unit	5A	4A
Particle size H/D	mm	3/1.5	3/1.5
Bulk density	kg/m^3	800	750
Water adsorption capacity	% (w/w)	22.7	20.4
pH stability	pH	5–12	5–12
Pore cross-section	Å	5	4.2

The drying of the isoprene was performed in a device in which adsorption studies were possible under dynamic conditions.

The experiments were performed in columns of 20 mm diameter and 700 mm height packed with 100 g 4A and 5A molecular sieves, respectively. The runs were made at normal pressure and temperature with a volumetric flow rate of 500 cm^3 isoprene/100 g sieves·hr.

The refractive index of the isoprene of over 99 % purity was $n_D^{20} = 1.4220$ before passing through the molecular sieves and that of the *ca.* 25 % pure isoprene fraction was $n_D^{20} = 1.3721$; the moisture content was about 40 ppm in both isoprene samples. The experimental results are given in Table 7.

Table 7

Drying of >99 % and ~25 % isoprene on 4A and 5A molecular sieves at a volumetric flow rate of 500 cm^3 isoprene/100 g sieves · hr

Molecular sieves	Moisture, ppm	
	Before drying	After drying
	>99 % isoprene	
4A	40	4
4A	40	3
5A	40	5
5A	40	4
	~25 % isoprene	
4A	38	4
4A	38	3
5A	38	5
5A	38	3

According to the data shown in Table 7, the refractive index of isoprene was not changed after passing through 4A or 5A molecular sieves. Since isoprene is a branched diene with a critical cross-section of the molecule of 5.9 Å, it is not retained within the pores of the 4A or 5A molecular sieves.

Drying below a level of 5 ppm water was performed by passage of both isoprene samples through 4A or 5A molecular sieves. The critical cross-section of water molecules is 3.2 Å, which enables a good drying on 4A molecular sieves, as shown by the experimental results.

The optimum volumetric flow rate for the drying of both isoprene samples on 4A molecular sieves was also verified. The volumetric flow rate was varied between 500 and 4000 cm^3 isoprene/100 g sieves·hr. The results are listed in Table 8.

Table 8
Effect of volumetric flow rate on capacity of 4A molecular sieves for water adsorption

Volumetric flow rate cm^3 isoprene/100 g sieve·hr	Moisture, ppm	
	Before purification	After purification
500	40	3
1000	40	4
1500	40	3
2000	40	5
3000	40	9
4000	40	11

The conclusion is that for volumetric flow rates up to 2000 cm^3 isoprene/100 g sieves·hr a good drying of both isoprene samples is achieved. At higher flow rates, the moisture content exceeds the limits permitted for polymerization.

4. *Removal of carbonyl compounds from isoprene*

The removal of carbonyl compounds from both samples of isoprene was carried out by using molecular sieves. The method is based on the co-adsorption of water and other polar impurities, such as carbonyl compounds, which, owing to their small critical cross-section, can be retained within the channels of 4A or 5A molecular sieves.

The adsorption of carbonyl compounds was studied under dynamic conditions in the same apparatus used for the removal of moisture from isoprene, at normal pressure and temperature and with a constant volumetric flow rate. The original carbonyl compound content of the two isoprene samples ranged from 50 to 300 ppm.

The retention efficiency of 4A and 5A molecular sieves was tested by measuring the concentration of carbonyl compounds by means of a photocolorimetric method. The method involves the colorimetric determination, in alkaline medium, of the hydrazone formed by reaction of the carbonyl compounds with 2,4-dinitrophenylhydrazine [389]. The results are shown in Table 9.

The experimental data have led to the conclusion that 5A molecular sieves remove carbonyl compounds from isoprene down to levels below 10 ppm. The adsorption capacity of 4A molecular sieves for carbonyl compounds is smaller since the critical cross-section of the carbonyl compounds is larger than the channel cross-section.

Table 9

Removal of carbonyl compounds from >99% and ~25% isoprene (volumetric flow rate of isoprene 100 cm^3/100 g sieves · hr)

Molecular sieves	Carbonyl compounds, ppm	
	Before purification	After purification
	>99% isoprene	
4A	50	20
4A	300	45
5A	50	7
5A	300	9
	~25% isoprene	
4A	55	25
4A	280	40
5A	55	8
5A	280	10

The effect of the volumetric flow rate on the retention capacity of 5A molecular sieves for carbonyl compounds has also been studied. The volumetric flow rate varied from 100 to 400 cm^3 isoprene/100 g sieves·hr. The isoprene samples had an original content of 300 ppm carbonyl compounds. The experimental results are given in Table 10. It shows that for volumetric flow rates up to 300 cm^3 isoprene/100 g sieves·hr the concentration of carbonyl compounds is below 10 ppm. At higher volumetric flow rates, and shorter contact times, the retention of carbonyl compounds is reduced.

Table 10

Effect of volumetric flow rate on adsorption capacity of 5A molecular sieves for carbonyl compounds

Volumetric flow rate cm^3 isoprene/100 g of sieve · hr	Carbonyl compounds (ppm)	
	Before purification	After purification
100	300	5
200	300	7
300	300	9
400	300	55

5. *Removal of acetylenic hydrocarbons from isoprene*

Removal of acetylenic hydrocarbons from isoprene was achieved with 4A or 5A molecular sieves. The apparatus and the working conditions were similar to those used for the removal of moisture and carbonyl compounds.

The content of acetylenic hydrocarbons in isoprene was determined by reaction with silver nitrate [390]. The results are listed in Table 11 at a volumetric flow rate of 100 cm^3 isoprene/100 g sieves·hr.

If isoprene is passed through 5A molecular sieves, the concentration of acetylenic hydrocarbons is reduced below 10 ppm. The 4A molecular sieves are not efficient, as their channels have a smaller cross-section than the critical cross-section of the acetylenic hydrocarbons.

The data on the capacity of 5A molecular sieves to retain acetylenic hydrocarbons from isoprene at volumetric flow rates ranging from 100 to 300 cm^3 isoprene/100 g sieves·hr are shown in Table 12.

The efficiency of 5A molecular sieves is related to the volumetric flow rate. It was established that volumetric flow rates higher than 200 cm^3 isoprene/100 g sieves·hr cannot be used.

Table 11

Removal of acetylenic hydrocarbons from over 99 % and *ca.* 25 % isoprene fractions; volumetric flow rate 100 cm^3 isoprene/100 g sieve · hr

Molecular sieves	Acetylenic hydrocarbons (ppm)	
	Before purification	After purification
	over 99 % isoprene	
4A	30	25
4A	30	22
5A	30	8
5A	30	6
	ca. 25 % isoprene	
4A	28	25
4A	28	20
5A	28	8
5A	28	5

Table 12

Effect of volumetric flow rate on adsorption capacity of 5A molecular sieves for acetylenic hydrocarbons

Volumetric flow rate cm^3 isoprene/100 g sieve · hr	Acetylenic hydrocarbons (ppm)	
	Before purification	After purification
100	30	5
200	30	8
300	30	15

From experiments on isoprene purification with molecular sieves, the following conclusions can be drawn:

— the water content of isoprene is reduced from 40 ppm to below 10 ppm by flowing through 4A molecular sieves at a volumetric flow rate of 2000 cm^3 isoprene/100 g sieves·hr. In such conditions, 100-g sieves can be used to purify about 600 litres of isoprene;

— the presence of carbonyl compounds or of acetylenic hydrocarbons in concentrations higher than 10 ppm (about 300 ppm of carbonyl compounds and 30 ppm of acetylenic hydrocarbons) involves the concomitant purification of isoprene from water, carbonyl compounds and acetylenes by passing through 5A molecular sieves at a volumetric flow rate of 200 cm^3 isoprene/100 g sieves·hr. The presence of acetylenic hydrocarbons necessitates this flow rate.

B. Solvent purification

Stereospecific polymerization of isoprene of over 99 % purity takes place in solution, using a saturated aliphatic hydrocarbon as reaction medium [391, 392], at a solvent/monomer ratio ranging from 3:1 to 5:1 (v/v) [393].

The purity requirements for the solvent are of the same order as for the monomer and a removal of impurities is therefore necessary [394].

Solvents used in the polymerization experiments were *n*-heptane and a hexane fraction with the following physical constants:

n-heptane $n_D^{20} = 1.3879$

$d_4^{20} = 684\ kg/m^3$

b.p. = 98.2–98.4°C at 760 mmHg

hexane fraction $n_D^{20} = 1.3765$

$d_4^{20} = 663\ kg/m^3$

b.p. range: 65–72°C at 760 mmHg

The chemical composition of the solvents as determined by GLC is given in Table 13 and shown in the chromatograms (Figures 3 and 4).

Chromatographic analyses show that the concentration of *n*-heptane was 98.72 % and that the hexane fraction contained 88.67 % *n*- and *iso*-hexane in about equal amounts. The remaining components of the above solvents need not be removed since they belong to the same class of hydrocarbons and do not affect the course of the polymerization reaction.

The solvents used have also been analysed by chemical methods in order to determine the moisture and oxygen content. The same procedures were

Table 13
GLC analysis of *n*-heptane and hexane fraction

Compounds	*n*-Heptane (%)	Hexane fraction (%)
Isopentane		0.36
n-Pentane	0.0035	0.70
i-Hexane	0.0001	1.20
	0.0004	20.90
		21.05
*n*Hexane	0.039	45.52
i-Heptane	0.33	1.80
	0.85	7.00
		0.09
		0.25
		0.70
n-Heptane	98.72	0.06
i-Octane	0.013	—
	0.0085	—
	0.006	—
n-Octane	0.015	—
Benzene	—	0.37
Toluene	0.01	0.003

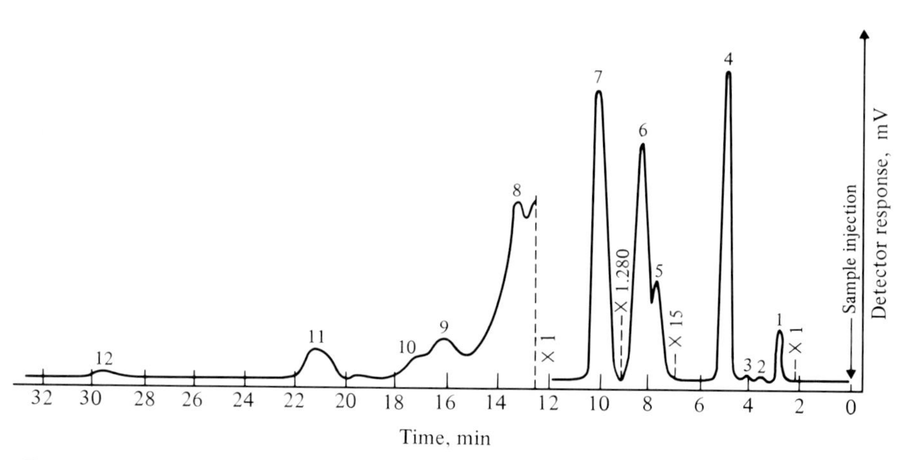

FIGURE 3. Chromatogram of *n*-heptane. 1—*n*-pentane; 2, 3—*iso*-hexane; 4—*n*-hexane; 5, 6—*iso*-heptane; 7—*n*-heptane; 8, 9, 10—*iso*-octane; 11—*n*-octane; 12—toluene.

applied as for isoprene. It was concluded that both *n*-heptane and the hexane fraction contained 20–30 ppm oxygen and 30–60 ppm water, respectively.

Oxygen was removed by distillation and water by adsorption on silica gel or on 4A molecular sieves to a level below 10 ppm, by the same procedures used in the purification of isoprene.

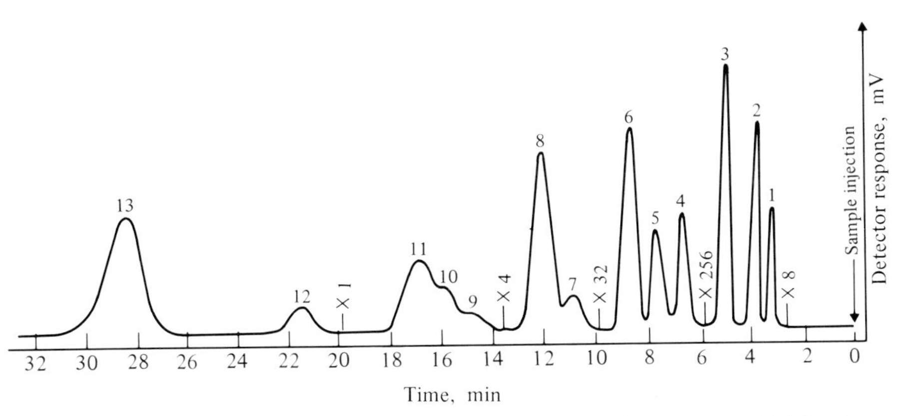

FIGURE 4. Chromatogram of hexane fraction. 1—isopentane; 2—*n*-pentane; 3, 4, 5—*iso*-hexane; 6—*n*-hexane; 7–11—*iso*-heptane; 12—*n*-heptane; 13—benzene.

1. Removal of oxygen from the n-*heptane and the hexane fraction*

Oxygen removal (deaeration) from the solvents used in polymerization was achieved by distillation on a 5 TP column under a stream of purified nitrogen. *n*-Heptane was collected at 98.2–98.4°C at 760 mmHg and the hexane fraction at 65–72°C at 760 mmHg.

The analytical data show that after distillation the oxygen content decreases from 20–30 to 2–5 ppm.

It was therefore concluded that the above method enables oxygen to be removed to a level below the accepted limit. Oxygen analysis was performed by the same Hersch's electrochemical method [383].

2. Removal of moisture from solvent

Removal of water from the solvents was achieved by adsorption on silica gel or on molecular sieves.

(*a*) *Moisture removal with silica gel.* To remove the moisture from both *n*-heptane and the hexane fraction, the activated silica gel had the same characteristics as the one used in the purification of isoprene. Purification of the solvent was effected after distillation by storing it over various amounts of silica gel for different intervals of time.

The experimental data indicate that by storing on 1.5 % activated silica gel for 1 hour, the water content of *n*-heptane and of the hexane fraction is reduced to about 10 ppm, while the basic chemical composition remains unchanged.

(*b*) *Moisture removal with molecular sieves.* Molecular sieves with a higher adsorption capacity and selective separation capacity were used for drying the solvents.

4A molecular sieves were tested for drying *n*-heptane and the hexane fraction. As it is known, *n*-paraffins have a molecular critical cross-section of 4.9 Å and cannot be retained on such sieves. Only components with a lower critical cross-section can be adsorbed, such as water whose critical cross-section is only 3.2 Å.

The characteristics of the molecular sieves used, the experimental apparatus and the working techniques were the same as those used for the purification of isoprene.

The drying of *n*-heptane and of the hexane fraction was studied as a function of the volumetric flow rate. The results are given in Table 14.

The capacity of the molecular sieves to retain water at high volumetric flow rates was determined. Good drying, to levels of water content below 10 ppm, was achieved at a volumetric flow rate of 4000 cm^3 solvent/100 g sieves·hr. About 450 litres solvent containing *ca.* 60 ppm water can be dried on 100 g 4A molecular sieves.

Table 14

Effect of volumetric flow rates on adsorption capacity of 4A molecular sieves for water from solvents

Volumetric flow rate cm^3/100 g sieve · hr	Moisture (ppm)			
	n-Heptane		Hexane fraction	
	Before drying	After drying	Before drying	After drying
1000	40	4	50	3
2000	40	3	50	4
3000	40	4	50	5
4000	40	7	50	8
5000	40	11	50	12
6000	40	20	50	22

C. Purification of catalyst components

For polymerization experiments, a catalyst system involving triethylaluminium or triisobutylaluminium and titanium tetrachloride was used.

The aluminium alkyls are obtained by direct synthesis from aluminium, hydrogen and the corresponding olefin [395, 396].

Since the reactivity of aluminium alkyl is reduced in the presence of traces of oxygen and moisture, such compounds are handled only under purified nitrogen.

Triethylaluminium (TEA) and triisobutylaluminium (TIBA) have the following physical constants, respectively [396]:

	TEA	TIBA
Formula	$Al(C_2H_5)_3$	$Al(i\text{-}C_4H_9)_3$
Molecular weight	114.15	198.31
Boiling point	70°C at 2 mmHg	47°C at 1 mmHg
Density	$840\ kg/m^3$ at 20°C	$786\ kg/m^3$ at 20°C
Refractive index	1.480 at 6.5°C	1.449 at 20°C
Solubility	soluble in hydrocarbons	soluble in hydrocarbons
Thermal stability	at higher temperatures (*ca.* 150°C) it slowly decomposes to diethylaluminium hydride and ethylene	at a temperature of about 70°C it decomposes to diisobutylaluminium hydride and isobutylene

The materials used were clear liquids with the following characteristics [396]:

	TEA	TIBA
Colour	colourless	colourless
Al content	*ca.* 23 %	*ca.* 13.5 %
Activity	*ca.* 95 %	*ca.* 95 %
Molar composition	al-C_2H_5 *ca.* 90 %	al-i-C_4H_9 *ca.* 95 %
(al = $A\frac{1}{3}$)	al-H 2–6 %	al-n-C_4H_9 0–2 %
	al-C_4H_9 4–8 %	al-H 2–5 %

The activity of the aluminium alkyls, which is their main characteristic, is determined by the ratio between the active aluminium and the total content of aluminium [397, 398].

Active aluminium is the amount of aluminium present as trialkylaluminium and dialkylaluminium hydride [399]. In order to determine the active aluminium, a method was used which was based on the ability of the trialkylaluminium to reduce quantitatively the titanium tetrachloride to trichloride under the given conditions.

The resulting titanium trichloride is further oxidized with ferric sulphate and the ferrous sulphate formed is estimated with an oxidizing agent, e.g. potassium permanganate [397].

The following reactions occur:

$$AlR_3 + TiCl_4 \rightarrow AlR_2Cl + RTiCl_3$$

$$RTiCl_2 \rightarrow R\cdot + TiCl_3$$

$$Ti^{3+} + Fe^{3+} \rightarrow Ti^{4+} + Fe^{2+}$$

$$2KMnO_4 + 10FeSO_4 + 8H_2SO_4 \rightarrow K_2SO_4 + 5Fe_2(SO_4)_3 + 2MnSO_4 + 8H_2O$$

The content of active aluminium is calculated from the quantity of potassium permanganate consumed in the analysis.

Total aluminium was determined by an indirect complexometric method [398]. We obtained 23.11% Al in triethylaluminium and 13.4% Al in triisobutylaluminium. Analytical data indicated that the activity of triethylaluminium was 94.5% whilst that of the triisobutylaluminium was 94.8%. Since their activity was similar to data reported in literature, the aluminium alkyls were used in polymerization reactions without further purification.

The second component of the catalyst system used in the stereospecific polymerization of isoprene was titanium tetrachloride, a colourless liquid which fumes in air, being easily hydrolysed to titanium oxyhydrates, titanium oxychloride and hydrogen chloride [400].

The compound has the following physical properties [401]:

Boiling point	135.8°C at 760 mmHg
Melting point	−24.8°C
Density	1,730 kg/m^3 at 20°C

Literature data report that titanium tetrachloride can be used in a Ziegler–Natta-type catalyst complex as such, once or twice distilled at normal pressure or under high vacuum, or distilled over copper dust [402–407].

Titanium tetrachloride was prepared by the direct chlorination of titanium dioxide with chlorine gas in the presence of charcoal as a reducing agent [407]:

$$TiO_2 + 2Cl_2 + 2C = TiCl_4 + 2CO$$

The crude product was distilled in order to separate certain solid impurities (carbon, solid oxychlorides, and ferric chloride). Purification was effected on a column packed with copper first at total reflux, then collecting the 135–137°C fraction. Before use in polymerization reactions, the compound was redistilled under purified nitrogen through a fractionation column packed with copper gauze rings with 1600 mesh/cm^2. A colourless liquid with a boiling temperature of 135.8°C and d_4^{20} = 1,730 kg/m^3 was obtained.

The physical properties of the product were identical with those reported in literature. Iron, within 0.01–0.001%, was found by UV emission spectroscopy.

D. Purification of nitrogen

The stereospecific polymerization of isoprene, all reagent handling and the processing of the solutions resulting from polymerization are carried out under a purified inert gas. The nitrogen used must contain oxygen and water below 10 ppm.

The inert gas used was nitrogen. Its initial concentration was 99.5% containing 0.2–0.5% O_2 and about 800 ppm of water.

Several procedures for nitrogen purification were tried in order to reduce the water and oxygen content below the given limits.

1. Removal of oxygen from nitrogen

Several methods for the removal of oxygen, classified into wet and dry processes, are reported in the literature [408–410].

Two dry purification processes, one with manganese oxide and one with a copper catalyst, were tested.

(*a*) *Removal of oxygen from nitrogen with manganese oxide.* This purification method is based on the oxidation reaction of manganese oxide with the oxygen present in the nitrogen; this results in the formation of manganese dioxide, the reaction taking place at room temperature and atmospheric pressure:

$$2MnO + O_2 = 2MnO_2$$

The device used to remove the oxygen from the nitrogen consisted of a 1000-mm column of 60 mm diameter, packed with manganese oxide beads of 3×3 mm in size.

After passing through the manganese oxide column at a flow rate of 30–40 litres/hr the issuing nitrogen gas contained about 10 ppm oxygen, as determined by Hersch's electrochemical method. At a higher flow rate, oxygen retention by the nitrogen gas was not satisfactory, as it exceeded the limit permitted for stereospecific polymerization.

Exhaustion of the column is monitored by analysis of the oxygen present in the nitrogen effluent after passage through the column. Column regeneration is carried out with hydrogen at 400–450°C to reduce the manganese dioxide:

$$MnO_2 + H_2 = MnO + H_2O$$

This method has the disadvantages that it does not ensure good purification at flow rates higher than 40 litres/hr, it is rapidly exhausted and it requires high temperatures for regeneration.

(*b*) *Removal of oxygen from nitrogen with a copper catalyst.* In order to achieve an improved purification of nitrogen at higher flow rates, a copper catalyst supported on magnesium silicate was tested. As stated in the literature, this catalyst can be used to remove various organic or inorganic impurities from liquids or gases. For instance, O_2, H_2, CO, acetylene or sulphur compounds can be removed from various gases either by catalytic conversions or by adsorption. The gases which can be thus purified are: nitrogen, carbon dioxide, saturated or unsaturated gaseous hydrocarbons, etc. [410].

The removal of oxygen from nitrogen with the copper catalyst is based on the following oxidation reaction of copper:

$$2Cu + O_2 = 2CuO$$

The catalyst contains 30% copper supported on magnesium silicate with various group VIII metals as additives.

The catalyst characteristics are the following:

Size	5 × 5 mm
Bulk density	750–850 g/m^3
Compression strength	*ca.* 200 kgf/cm^2
Specific heat	0.264 kcal/kg°C
Working temperature	0–250°C
Working pressure	up to 300 kgf/cm^2

Before use the copper catalyst must be activated since it is in an oxidized state. Activation is carried out by reduction with hydrogen at 150–180°C, which converts the copper oxide into copper metal, according to the equation:

$$CuO + H_2 = Cu + H_2O$$

The reaction is exothermic and the temperature is maintained below 180°C by controlling the hydrogen flow. Temperatures higher than 250°C destroy the catalyst. Completion of the activation reaction is observed when water no longer condenses on the walls of the column. The column is afterwards cooled to room temperature under nitrogen.

Removal of oxygen from nitrogen with a copper catalyst was carried out in an equipment consisting of a metallic column of 70 mm diameter and 1500 mm height, containing about 4500 g catalyst. At a nitrogen flow rate of 60–100 litres/hr, the oxygen content as determined by Hersch's electrochemical method, was reduced below 5 ppm.

About 50–60 m^3 nitrogen can be purified with such a column packed with copper catalyst; afterwards it must be regenerated. Regeneration effected is in the same way as activation. The copper catalyst has a high mechanical strength and can be regenerated several times without loss of activity.

Nitrogen purification from oxygen with copper catalyst offers a number of advantages: it enables oxygen removal at high flow rates, both activation and regeneration are effected at lower temperatures than in the case of manganese oxide, and the catalyst can be regenerated several times owing to its high mechanical strength.

2. Removal of moisture from nitrogen

The original water content of the nitrogen was 800 ppm. Given such a high degree of moisture, a stepwise drying procedure was applied which allowed a progressive removal of water, by passing the gas through a drying assembly of several columns containing drying agents in an increasing order of efficiency.

The usual drying agents have been used. Their relative efficiency is given in Table 15.

Table 15
Relative efficiency of usual drying agents [386]

Drying agent	Residual moisture mg H_2O/litre gas
Calcium chloride	0.36
Sodium hydroxide	0.16
Activated silica gel	0.006
Molecular sieves	0.00015
Phosphorus pentoxide	0.00002

The device used in the laboratory experiments for the removal of water from nitrogen consisted of four columns of 70 mm diameter and 1000 mm height containing 2000 g of calcium chloride, 2500 g of sodium hydroxide, 2500 g of silica gel, 800 g of phosphorus pentoxide, or 1000 g of 4A molecular sieve.

With this drying system, an almost complete removal of water, below *ca.* 5 ppm is achieved as determined by the Karl Fischer method. About 30–40 m^3 of nitrogen, at a flow rate of 50–60 litres/hr, could be purified by the above-mentioned device.

Discussion on the purification of reagents. The investigation of the purification of reagents and inert gas used in the stereospcific polymerization of isoprene was necessary since they contained a number of components interfering with the polymerization process whose level had therefore to be reduced below certain allowed limits.

The initial analyses of the reagents indicated that they contained certain specific impurities, such as cyclopentadiene, carbonyl compounds and acetylenic hydrocarbons in isoprene. Oxygen and water are present in monomer, solvent and nitrogen.

The purification method is primarily dependent on the nature of the impurity. The same method could, however, be used to remove a particular impurity present in several reagents.

Purification procedures were chosen such as to achieve the concomitant removal of several components where present together, in order to reduce the number of operations required.

The impurities present in the reagents were removed as follows:

— cyclopentadiene, by reaction with maleic anhydride;
— oxygen from isoprene and solvents, by distillation, and from nitrogen, with a copper catalyst;
— water, with sorbents, better results being obtained with molecular sieves, than with silica gel;
— carbonyl compounds and acetylenic hydrocarbons present in isoprene were removed by passing through molecular sieves.

The purification of reagents and inert gas was thus performed by chemical reactions, distillation and percolation procedures through molecular sieves, the latter allowed the simultaneous retention of three impurities owing to their co-adsorption properties.

The above investigation led to the conclusion that the chosen purification procedures are efficient as they provide an almost complete removal of the impurities present in the reagents.

Chapter II

Research on the stereospecific polymerization of isoprene with heterogeneous catalyst systems of the $AlR_3 + TiCl_4$ type

THE main aim of the research on the stereospecific polymerization of isoprene was to obtain a polyisoprene with more than 92% *cis*-1,4 structural units and with physical-mechanical properties similar to natural rubber.

As it is known, owing to the asymmetry of the isoprene molecule, four routes are possible for the addition of the monomeric units to a growing chain:

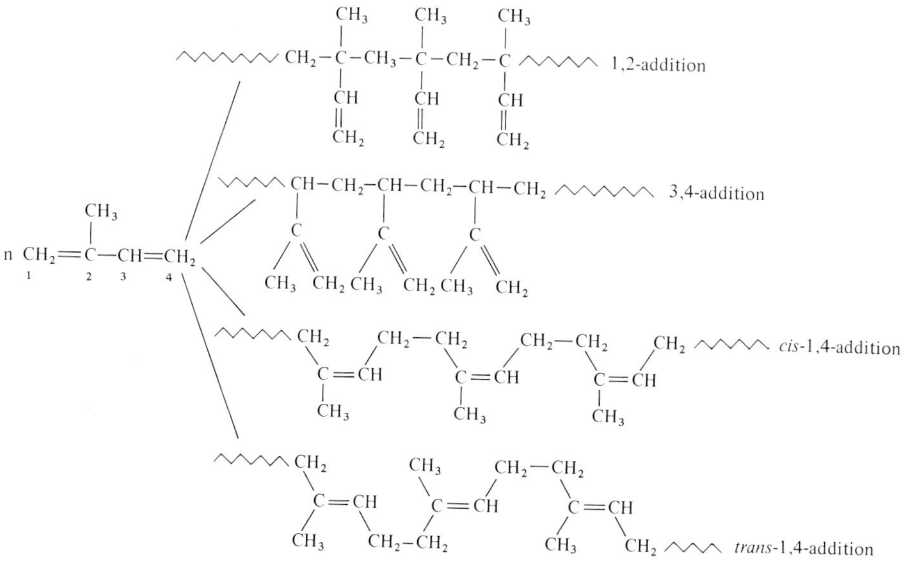

```
                              CH3        CH3         CH3
                               |          |           |
                ~~~~~ CH2—C—CH2—C—CH2—C ~~~~~  1,2-addition
                               |          |           |
                              CH         CH          CH
                              ||         ||          ||
                              CH2        CH2         CH2

                 ~~~~ CH—CH2—CH—CH2—CH—CH2 ~~~~~  3,4-addition
                       |         |         |
     CH3               C         C         C
      |               / \\      / \\      / \\
n CH2=C—CH=CH2      CH3  CH2  CH3  CH2  CH3  CH2
  1   2  3  4

                 ~~~~ CH2        CH2—CH2         CH2—CH2        CH2 ~~~~ cis-1,4-addition
                         \      /       \       /       \      /
                          C=CH           C=CH            C=CH
                          |              |               |
                          CH3            CH3             CH3

                 ~~~~ CH2           CH3       CH2—CH2
                         \             \      /       \
                          C=CH          C=CH           C=CH
                         /    \        /              /    \
                      CH3     CH2—CH2             CH3      CH2 ~~~~ trans-1,4-addition
```

In 1,4-addition, both double bonds of the isoprene molecule are converted into single bonds, while the central bond is transformed into a double bond; hence, the double bond of the in-chain unit is placed in a different position from those in the initial isoprene molecule. The two carbon atoms, C2 and C3, involved in the double bond of the 1,4-addition isomer bear three different types of substituents. The free rotation of these two carbon atoms is hindered by the double bond and the monomer unit can therefore take two conformations:

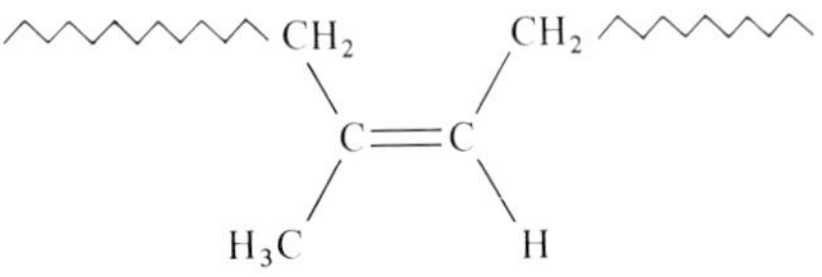

```
~~~~~CH2          CH2~~~~~          ~~~~~CH2        H
        \        /                          \      /
         C======C                            C====C
        /        \                          /      \
     H3C          H                      H3C        CH2~~~~~

          cis                                  trans
```

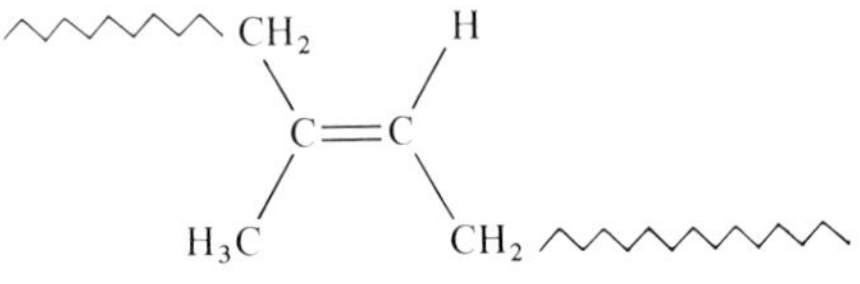

Polymers with such structures are geometrical isomers and their physical properties are markedly different. The geometrical isomer with monomeric units added in a *cis*-1,4 position is similar to the structure of natural rubber.

cis-1,4-Polyisoprene is synthetically prepared by the polymerization of isoprene with stereospecific catalysts via an anionic coordination reaction mechanism.

In this work, the conditions of preparation of a *cis*-1,4-polyisoprene with improved elastomeric properties was investigated. Elastomeric properties depend on the degree of stereoregularity and on the molecular weight of the polymer. Polymer stereoregularity is determined by the stereospecificity of the catalyst system, which in its turn depends on the nature of the catalyst components, their molar ratio and the purity of the reaction medium. Molecular weight, another important property, also depends on the catalyst system and on temperature.

As already mentioned, in the present research work, two isoprene fractions with purities of over 99% and *ca.* 25% were used, together with two catalyst systems of the $AlR_3 + TiCl_4$ type and a saturated aliphatic hydrocarbon as solvent.

The influence of the nature of the catalyst system, the molar ratio of the catalyst components and the catalyst concentration were studied because of the important role played by the catalyst in the preparation of polymers with elastomeric properties.

The catalyst systems studied included triethylaluminium and triisobutylaluminium, respectively, complexed with titanium tetrachloride. For both catalyst systems, the influence of the Al/Ti molar ratio and of the catalyst concentration have been investigated since they control both the polymer microstructure and the molecular weight, as well as the rate of the polymerization reaction. The influence of monomer purity, of the nature of the solvent, of the reaction temperature and of the polymerization time on polymer properties, as well as the mechanism and kinetics of the polymerization reaction, have also been investigated.

The physical-chemical properties of the polymer obtained from each of the isoprene fractions by using the two catalyst systems have been investigated by determining their microstructure, molecular weight, crystallinity and degree of unsaturation. The physical-mechanical properties were determined by measuring the tensile strength, the modulus of elasticity, the elongation at break and the set at break.

A. Polymerization of isoprene of over 99 % purity

Two catalyst systems, namely triethylaluminium (TEA) + titanium tetrachloride and triisobutylaluminium (TIBA) + titanium tetrachloride, and *n*-heptane or a hexane fraction as solvents have been used in experiments to polymerize isoprene of over 99 % purity.

Both the Al/Ti molar ratio and the catalyst concentration have been varied with each of the catalyst systems. The influence of impurities on the polymerization reaction when using the TEA + $TiCl_4$ catalyst system and of the nature of the solvent when using the TIBA + $TiCl_4$ catalyst complex have also been investigated.

The polymerization reaction was effected at atmospheric pressure, under purified nitrogen. The catalyst complex was prepared *in situ* (within the monomer–solvent mixture). The reaction temperature was kept constant by removing the heat evolved.

After purification, the reagents used in polymerization had the following characteristics:

	Isoprene	*n*-Heptane	Hexane fraction
n_D^{20}	1.4220	1.3879	1.3765
d_4^{20}, kg/m^3	682	684	615
B.p. °C at 760 mmHg	34.1	98.2–98.4	65–72
Dielectric constant	2.02	1.924	1.890
	Triethylaluminium		Triisobutylaluminium
Activity, %	94.5		94.8
Total aluminium, %	23.11		13.41
	Titanium tetrachloride		
Concentration, %	99.9		
B.p. °C at 760 mmHg	135.8		
d_4^{20}, kg/m^3	1730		

1. Polymerization of isoprene of 99% purity with triethylaluminium and titanium tetrachloride

In polymerization experiments performed with the catalyst system prepared from triethylaluminium and titanium tetrachloride, using *n*-heptane as solvent, the influence of the following variables on the polymerization processes and on the polymer properties has been studied:

— Al/Ti molar ratio,
— catalyst concentration,
— monomer purity.

(*a*) *Influence of Al/Ti molar ratio.* The influence of the Al/Ti molar ratio on the polymer properties was examined in order to establish an optimum molar ratio for the preparation of polymers with high stereoregularity, conversion and molecular weight. The Al/Ti molar ratio varied between 0.5 and 2.0, keeping the other reaction conditions constant: monomer/solvent ratio = 1/4 (v/v), catalyst concentration = 1 g TEA/100 g isoprene, reaction temperature = 32–35°C, and reaction time = 120 minutes.

Characterization of the resultant polymers has been made by determining their microstructure, the viscosity molecular weight and the physical-mechanical properties. The results are listed in Table 16.

As shown in Table 16, both the polymer characteristics and the conversion differ markedly with the Al/Ti molar ratio used in the polymerization experiments.

At an Al/Ti molar ratio of 0.5, the polymers contained only 30% of *cis*-1,4-addition product and more than 50% of *trans*-1,4 units, as shown by the IR absorption spectrum given in Figure 5a. At this Al/Ti molar ratio, the polymers have a low molecular weight, as their intrinsic viscosity is about 0.2 dl/g. The polymerization reaction is slow and conversion is about 25%. The polymer has a granular appearance, a crosslinked structure, and contains over 30% of insoluble material; therefore, it cannot be used as an elastomer.

At Al/Ti molar ratios of 1.5 and 2, the polymers contain *ca.* 93% of *cis*-1,4 addition product, but the molecular weight and conversion are both low; the intrinsic viscosity was about 1 and 0.1 dl/g, respectively, and conversion was only 20% and 2%, respectively. Liquid polymers with low molecular weights were obtained.

In polymerizations using an Al/Ti molar ratio of 1, polymers with good elastomeric properties were obtained, with about 94% *cis*-1,4 addition product, as proved by the IR absorption spectrum shown in Figure 5b. Further, the molecular weight was high, the intrinsic viscosity having values between 4.5 and 5 dl/g, and the conversion reached 60%. The polymers also showed improved physical-mechanical properties, i.e. tensile strengths of 270 kgf/cm^2 and elongations at break of 700%.

Table 16
Effect of Al/Ti molar ratio

No.	Al/Ti molar ratio	Conversion (%)	[η] (dl/g)	Properties of polyisoprene: Microstructure				Physico-mechanical properties				Remarks
				cis-1,4 (%)	*trans*-1,4 (%)	3;4 (%)	1,2 (%)	Tensile strength (kgf/cm^2)	Elongation at break (%)	Modulus at 300% elongation (kgf/cm^2)	Set at break (%)	
1	0.5	20	0.21	27	58	9.5	5.5	—	—	—	—	Granular
2	0.5	28	0.18	31	53.5	9.5	6	—	—	—	—	polymer
3	0.5	25	0.20	30	54.5	10	5.5	—	—	—	—	
4	1.0	55	4.5	94	0	4	2	270	700	14	8	
5	1.0	60	5.0	93	0	4	3	280	690	16	8	
6	1.0	58	4.8	92.5	0	5	2.5	265	710	15	9	
7	1.2	53	4.4	93	0	4	3	265	710	17	8	
8	1.2	56	4.8	92.5	0	5	2.5	270	700	15	9	
9	1.2	51	4.1	94	0	4	2	266	720	18	8	
10	1.5	15	1.1	92	0	5	3	—	—	—	—	Oily
11	1.5	20	0.8	92.5	0	4	3.5	—	—	—	—	polymer
12	1.5	18	0.9	93	0	4	3	—	—	—	—	
13	2.0	1.8	0.1	92.5	0	4.5	3	—	—	—	—	Oily
14	2.0	2.5	0.15	93	0	4	3	—	—	—	—	polymer
15	2.0	2	0.1	94	0	4	2	—	—	—	—	

Over 99% isoprene fraction.
1 g TEA/100 g isoprene.

Polymers with similar properties were also obtained at an Al/Ti molar ratio of 1.2. However, a relative decrease in the molecular weight and in the conversion was noticeable, which did not affect the physical-mechanical properties.

The IR absorption spectra, recorded with a UR 10 Zeiss–Jena spectrophotometer, of the polymers obtained at various Al/Ti molar ratios are shown in Figures 5a, 5b, 5c.

The changes in microstructure, molecular weight and conversion with variation in the Al/Ti molar ratio are plotted in Figures 6, 7 and 8, respectively.

The spectra given in Figures 5a, b, c and the diagram shown in Figure 6 indicate the influence of the Al/Ti molar ratio on polymer microstructure. Maximum contents of the *cis*-1,4-units were obtained for Al/Ti molar ratios of 1 and higher than 1.

The plots drawn in Figures 7 and 8 show that both the molecular weight and the conversion reach maximum values for an Al/Ti molar ratio of 1.

As it is known, the most active, highly stereospecific, catalysts are formed by reaction of an alkyl derivative of an electropositive metal with a small ionic radius from groups I–III of the Periodic Table, and the halide of a transition metal from groups IV–VIII, the transition metal being in a lower valency state and having vacant *d*-orbitals. In the formation of the catalyst complex between the AlR_3 and the $TiCl_4$, the transition metal, tetravalent titanium, is reduced to a lower valency state, trivalent or divalent titanium, by the aluminium alkyl which plays the role of a reducing agent. The degree of reduction of the transition metal depends on the concentration of the aluminium alkyl in the catalyst system.

At an equimolar triethylaluminium/titanium tetrachloride ratio, an almost complete reduction of the transition metal from tetravalent to trivalent titanium takes place. The resulting complex shows the highest stereospecificity and activity, since reduced titanium, Ti(III) initiates isoprene polymerization. Experimental data confirm that at an Al/Ti molar ratio = 1, the catalyst complex shows the highest activity and selectivity, resulting in the formation of polymers with maximum reaction rate, stereoregularity and molecular weight.

In the catalyst complex formed with an Al/Ti molar ratio = 0.5, titanium is only partially reduced from its tetravalent state to trivalent titanium, because of an insufficient amount of alkylaluminium. The transition complex still contains titanium tetrachloride, i.e. non-reduced transition metal. Such a catalyst complex yields crosslinked polymers at a slow reaction rate and with low stereoregularity and molecular weight.

At an Al/Ti molar ratio = 2, the catalyst complex has a very low activity since the aluminium alkyl is present in a higher concentration and reduces the transition metal from Ti(IV) to Ti(II). As is well-known, the latter titanium valency state does not favour diene polymerization.

From the above data, it is concluded that the catalyst complex behaves

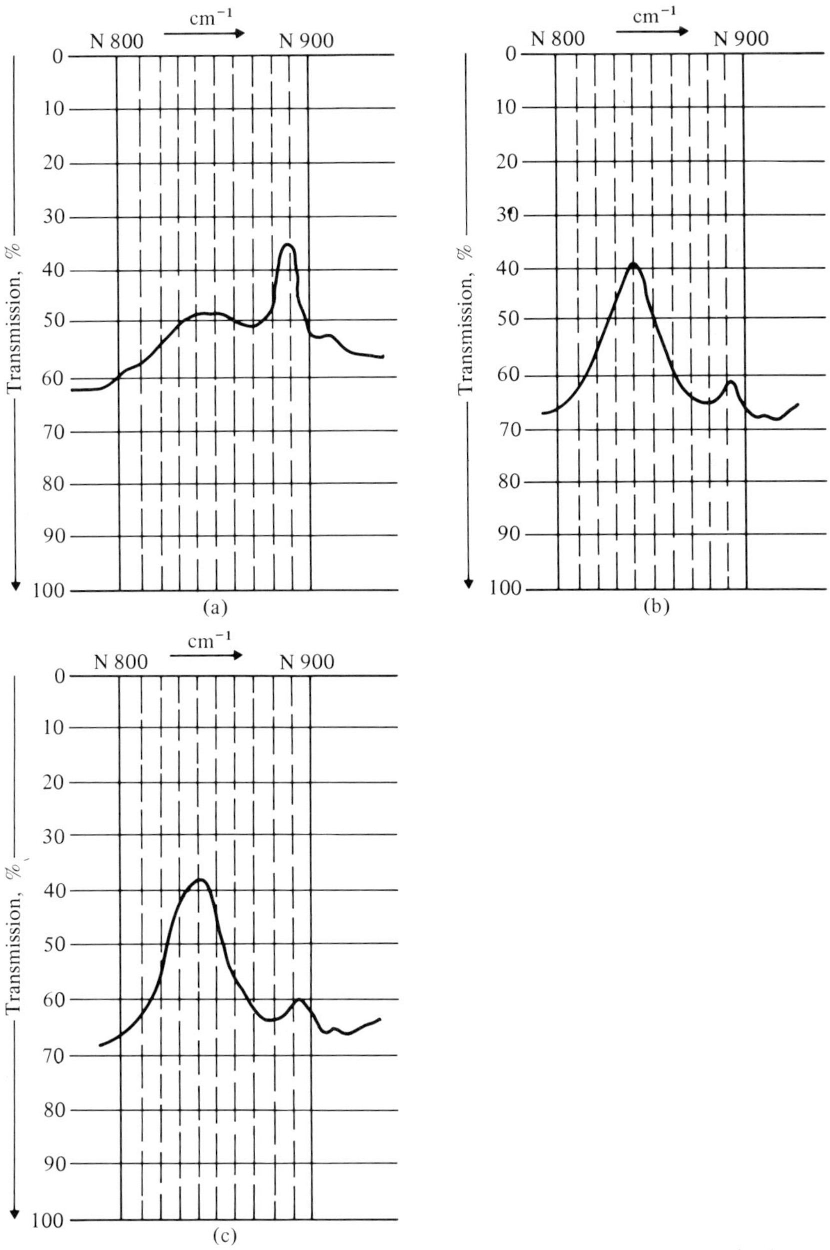

FIGURE 5. IR spectra of polyisoprene synthesized at various Al/Ti molar ratios, using isoprene fraction of over 99% purity. (a) Al/Ti molar ratio = 0.5. *cis*-1,4 = 31%; *trans*-1,4 = 53.5%; 3,4 = 9.5%; 1,2 = 6%. (b) Al/Ti molar ratio = 1. *cis*-1,4 = 94%; *trans*-1,4 = 0%; 3,4 = 4%; 1,2 = 2%. (c) Al/Ti molar ratio = 2. *cis*-1,4 = 92%; *trans*-1,4 = 0%; 3,4 = 4.5%; 1,2 = 3.5%.

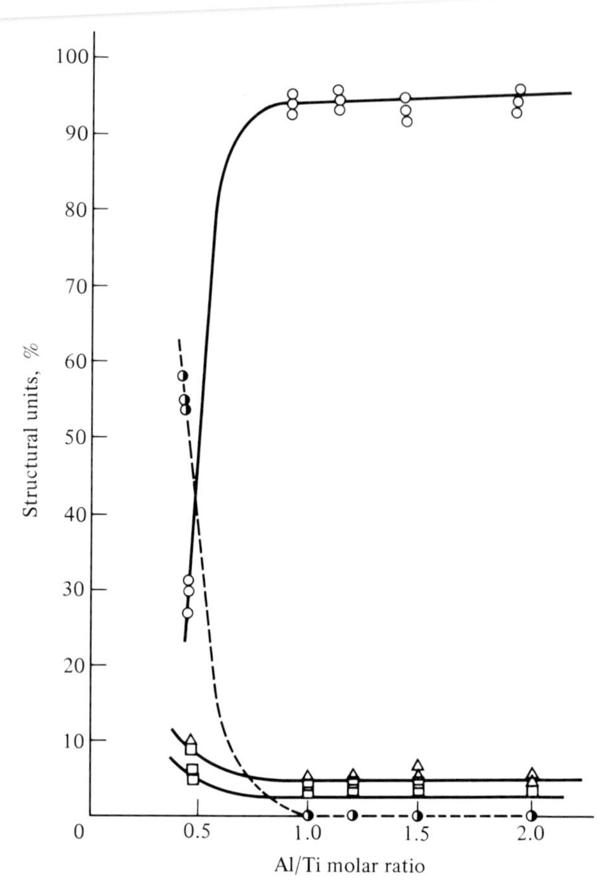

FIGURE 6. Effect of TEA/$TiCl_4$ molar ratio on microstructure of polyisoprene (isoprene fraction of over 99% purity). ○ *cis*-1,4; ◑ *trans*-1,4; △ 3,4; ☒ 1,2.

differently as a function of the Al/Ti molar ratio, whether it concerns the activity or the control of the monomer addition in the formation backbone chain of the polymer. The experimental data on the optimum Al/Ti molar ratio led to the conclusion that at a molar ratio of 1, polymers with high stereoregularity and molecular weight and with improved physico-mechanical properties are obtained, and a maximum reaction rate is reached as well.

(*b*) *Influence of concentration of catalyst.* The influence of the catalyst concentration on the characteristics of the polymer has been studied in order to find the best conditions for the synthesis of polymers with improved properties at convenient reaction rates.

The catalyst concentration (with respect to the monomer) was varied from 0.5 to 2.5 g TEA/100 g isoprene. An Al/Ti molar ratio = 1, yielding polymers with maximum stereoregularity, was used. The other variables, i.e. a monomer/solvent ratio of 1/4 (v/v), a reaction temperature of 32–35°C and a reaction time of 120 minutes, were kept constant. The prepared polymers were characterized by intrinsic viscosity, microstructure and physico-mechanical properties. The reaction rate was expressed by the final conversion. The results are given in Table 17.

From the results listed in Table 17, a correlation between catalyst concentration, molecular weight and reaction rate can be inferred.

As is well-known, the catalyst concentration influences the molecular weight; the higher the catalyst concentration, the lower the molecular weight. In runs using a catalyst concentration of 0.5 g TEA/100 g isoprene, polymers with an intrinsic viscosity of about 3 dl/g and conversions ranging from 42 to 46 % were obtained. The physico-mechanical properties of the polymers are relatively low, e.g. tensile strengths below 230 kgf/cm^2 and elongations at break, *ca.* 745 %.

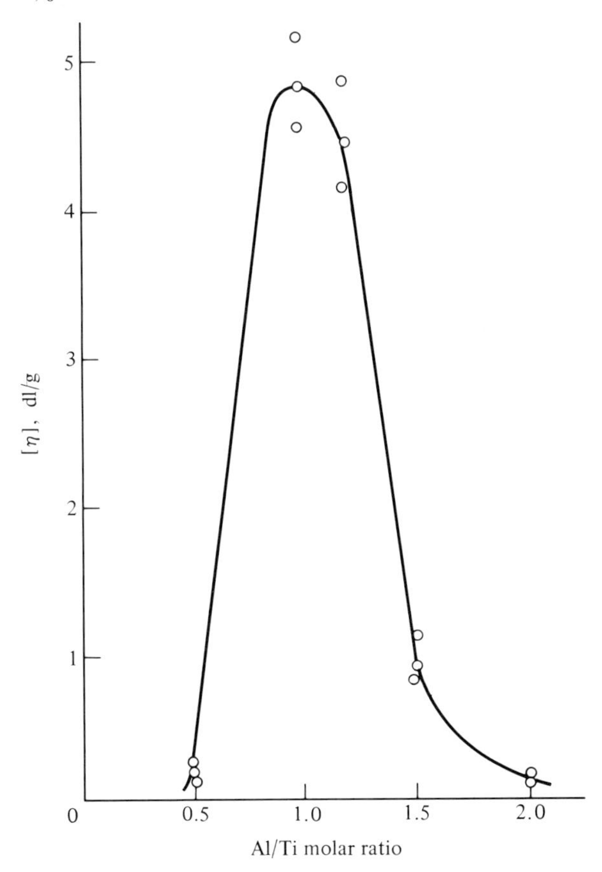

FIGURE 7. Effect of TEA/$TiCl_4$ molar ratio on intrinsic viscosity of polyisoprene (isoprene fraction of over 99 % purity).

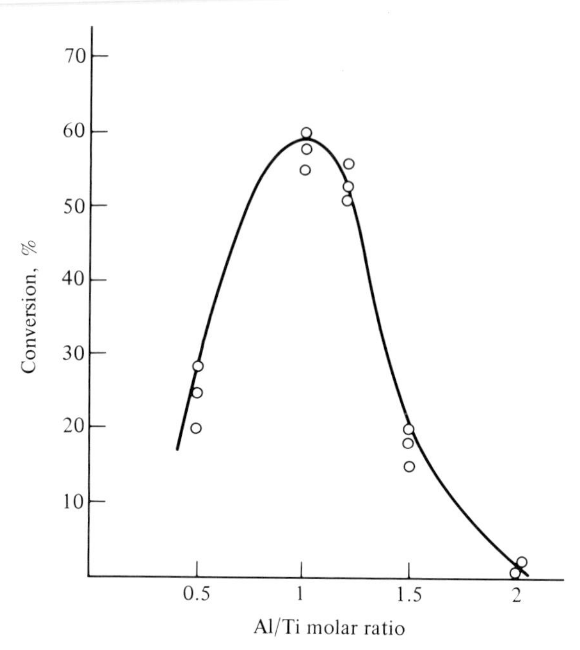

FIGURE 8. Effect of $TEA/TiCl_4$ molar ratio on conversion (isoprene fraction of over 99 % purity).

At a catalyst concentration of 1 g TEA/100 g isoprene, polymers with a high molecular weight (the intrinsic viscosity reaching 5 dl/g) and conversions up to 60 % were obtained. Polymers prepared in such a way had improved physico-mechanical properties: tensile strengths of 280 kgf/cm^2 and elongations at break of 700 %.

At a catalyst concentration of 1.5 g TEA/100 g isoprene, polymers with an intrinsic viscosity of *ca.* 4.2 dl/g were obtained and conversion ranged from 64 to 67 %. The physico-mechanical properties were also improved: e.g. tensile strengths of 270 kgf/cm^2 and elongations at break of 700 %.

The use of a catalyst concentration of 2 g TEA/100 g isoprene afforded intrinsic viscosities between 3 and 3.5 dl/g and conversion reached 74 %. The physico-mechanical properties were: tensile strengths of 240 kgf/cm^2 and elongations at break of over 730 %.

At a catalyst concentration of 2.5 g TEA/100 g isoprene, polymers with a lower molecular weight were obtained. The intrinsic viscosity was 2.5 to 2.8 dl/g. Higher reaction rates were noticed, conversion reaching 81 %. The physico-mechanical properties of the resulting polymers were: tensile strengths of 220 kgf/cm^2 and elongations at break, of *ca.* 750 %.

All polymers prepared with the above catalyst concentrations had a high content of the *cis*-1,4 addition product (more than 93 %); the catalyst concentration had no influence on the microstructure of the polymer.

Table 17

Effect of catalyst concentration

				Properties of polyisoprene							
				Microstructure				Physico-mechanical properties			
No.	g TEA/ 100 g isoprene	Con- version (%)	$[\eta]$ (dl/g)	*cis*-1,4 (%)	*trans*-1,4 (%)	3,4 (%)	1,2 (%)	Tensile strength (kfg/cm²)	Elong- ation at break (%)	Modulus at 300% elong- ation (kgf/cm²)	Set at break (%)
1	0.5	42	2.8	91.5	0	5.5	3	220	750	19	13
2	0.5	45	3.0	93	0	4.5	2.5	230	745	20	12
3	0.5	46	2.7	92	0	5	3	225	748	18	14
4	1.0	55	4.5	94	0	4	2	270	700	13	8
5	1.0	60	5.0	93	0	4	3	280	690	16	8
6	1.0	58	4.8	92.5	0	5	2.5	265	710	15	9
7	1.5	64	4.2	91.5	0	5.5	3	270	690	13	8
8	1.5	66	4.2	92.5	0	5	2.5	272	700	16	11
9	1.5	67	4.1	94	0	4	2	266	720	18	8
10	2.0	71	3.5	93	0	4.5	2.5	250	720	18	12
11	2.0	74	3.0	92	0	4.5	3.5	230	745	20	12
12	2.0	72	3.1	92.5	0	5	2.5	240	730	18	14
13	2.5	79	2.8	91.5	0	5.5	3	225	740	21	11
14	2.5	81	2.5	93	0	4.5	2.5	215	770	20	12
15	2.5	78	2.7	92	0	5	3	220	735	20	12

Over 99% isoprene fraction.
Al/Ti molar ratio = 1.

The changes of molecular weight and of reaction rate with the catalyst concentration are shown in Figures 9 and 10, respectively.

From the above experimental data, conclusions could be drawn as to the influence of the catalyst concentration on the molecular weight and reaction rate. Thus, the intrinsic viscosities obtained at various catalyst concentrations are plotted in Figure 9. The maximum value of the molecular weight is reached at a concentration of 1 g TEA/100 g isoprene. The molecular weight further decreases with increasing concentration of catalyst. At concentrations less than 1 g TEA/100 g isoprene, lower molecular weights resulted. The inversely proportional relationship between the molecular weight and the catalyst concentration holds only within the limits of 1 to 2.5 g TEA/100 g isoprene, and is not valid at a concentration of 0.5 g TEA/100 g isoprene. This result can be explained by the fact that in polymerizations with an *in situ* catalyst formation, the traces of impurities present in the reaction mixture at low catalyst concentrations, selectively destroy one of the catalyst components. The Al/Ti molar ratio is thus changed, resulting in the formation of low molecular weights.

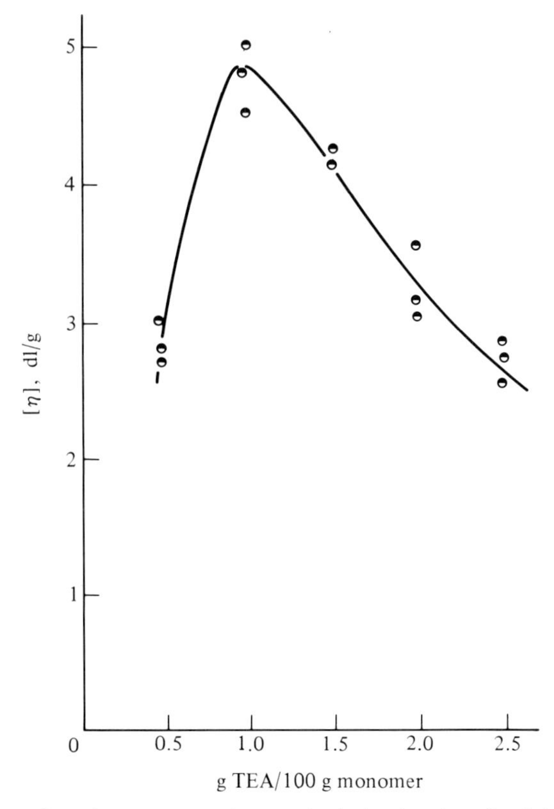

FIGURE 9. Effect of catalyst concentration on intrinsic viscosity of polyisoprene (isoprene fraction of over 99% purity).

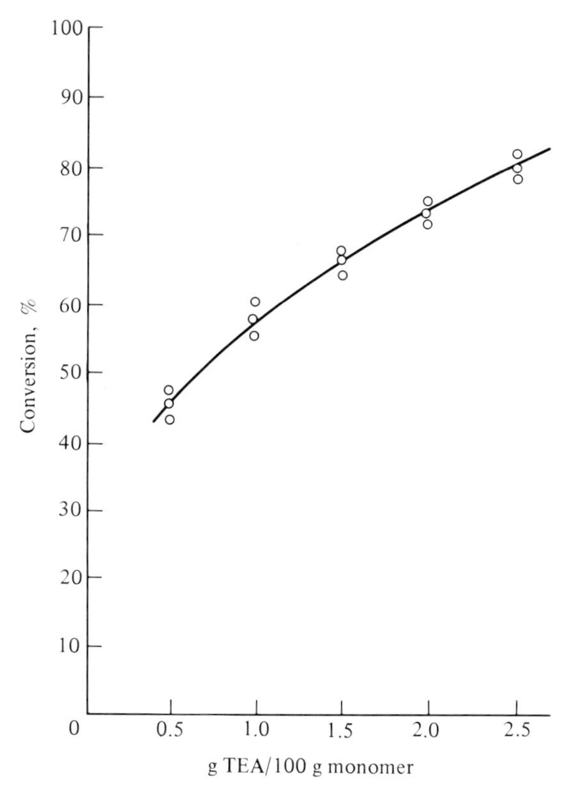

FIGURE 10. Effect of catalyst concentration on conversion (of isoprene fraction of over 99% purity).

The reaction rate, as expressed by the conversion, is proportional to the catalyst concentration, as shown in Figure 10. High conversions at high catalyst concentrations are related to the formation of numerous active sites which simultaneously initiate a great number of polymer chains with a low degree of polymerization and hence with a low molecular weight.

The characteristics of the polymers obtained at various catalyst concentrations indicate that in the concentration range from 1 to 1.5 g TEA/100 g isoprene, polymers are formed with intrinsic viscosities of 4 to 5 dl/g and improved physical-mechanical properties. Although the molecular weights show values over a relatively large range, the values of physico-mechanical properties are very narrow. This is explained by a more marked degradation of the higher molecular weight polymers during milling.

(*c*) *Influence of isoprene purity.* The variation of isoprene purity with time was investigated by watching the influence of storage time on polymer characteristics and on the polymerization exotherm. After distillation, isoprene was mixed with purified *n*-heptane in a 1:4 (v/v) ratio and stored under nitrogen at a temperature of 3–5°C. The mixture was used in polymerizations carried out after various storage times, namely 48, 96, 144 and 192 hours in reaction conditions affording optimum polymer properties, i.e. at an Al/Ti molar ratio of 1 and a catalyst concentration of 1 g TEA/100 g isoprene. The polymerization reaction was carried out under adiabatic conditions up to a temperature of 32–35°C, which was afterwards kept constant. The polymerization reaction time was 120 minutes. Characterization of the polymers was made by determining their microstructure, molecular weight and physico-mechanical properties. The results are listed in Table 18. The data emphasize that the storage time of the isoprene does not influence the polymer microstructure and molecular weight, but only the polymerization reaction rate. All data were compared with values obtained with polymers prepared without storing the monomer–solvent mixture.

The polymers resulting from the isoprene–*n*-heptane mixture which had been stored for different periods of time show a high stereoregularity, with a content of *cis*-1,4 addition product higher than 92 %. The molecular weight was also high, the intrinsic viscosity ranging from 4.3 to 5 dl/g. The physica-mechanical properties had high values, the tensile strength varying between 265 and 280 kgf/cm^2 and the elongation at break between 705 and 690 %.

The influence of storage time on the reaction rate becomes evident even after 48 hours, with conversion dropping to 50 % from the 60 % obtained when freshly distilled isoprene is polymerized. Conversion decreases continuously with increasing storage time of the isoprene–*n*-heptane mixture, reaching only 32 % after 192 hours.

The polymerization reaction exotherms for various times of storage are plotted in Figure 11. Temperature increase during the first stage of the polymerization reaction is also plotted to emphasize the effect on the induction period. Changes in conversion rate as a function of the storage time are shown in Figure 12. These diagrams show that after a 48-hour storage time of the isoprene–*n*-heptane mixture, conversion is about 10 % lower, although the exotherm is not very different from the one observed in the polymerization of the unstored isoprene. An increase in the storage time causes a longer induction period and a decrease in both the conversion and the reaction exotherm.

The above results show the need to purify the monomer–solvent mixture prior to polymerization, in order to avoid decreased reaction rate. The latter is probably caused by the formation in time of certain polymerization inhibitors which could not be detected by the analytical methods used.

2. Polymerization of isoprene of over 99 % purity with triisobutylaluminium and titanium tetrachloride

Investigation of the polymerization of isoprene of purity over 99 % with the catalyst system consisting of triisobutylaluminium and titanium tetrachloride

Table 18

Effect of storage time of the isoprene–*n*-heptane mixture

				Properties of polyisoprene							
				Microstructure				Physico-mechanical properties			
No.	Storage time (hr)	Conversion (%)	[η] (dl/g)	*cis*-1,4 (%)	*trans*-1,4 (%)	3,4 (%)	1,2 (%)	Tensile strength (kgf/cm^2)	Elongation at break (%)	Modulus at 300% elongation (kgf/cm^2)	Set at break (%)
1	0	55	4.5	94	0	4	2	270	700	14	8
2	0	60	5.0	93	0	4	3	280	690	16	8
3	48	50	4.6	92.5	0	5	2.5	265	705	16	9
4	48	48	4.4	93.5	0	4.5	2	266	700	17	8
5	96	44	4.5	92.5	0	4.5	3.5	272	712	15	9
6	96	45	4.7	93.5	0	4.5	2	275	710	15	8
7	144	39	4.4	92.5	0	5	2.5	265	700	17	9
8	144	37	4.3	92	0	4.5	3.5	268	690	16	10
9	192	32	4.5	93.5	0	4.5	2	270	710	16	10
10	192	34	4.3	92.5	0	5	2.5	265	695	17	9

Over 99% isoprene fraction.
Al/Ti molar ratio = 1.
1 g TEA/100 g isoprene.

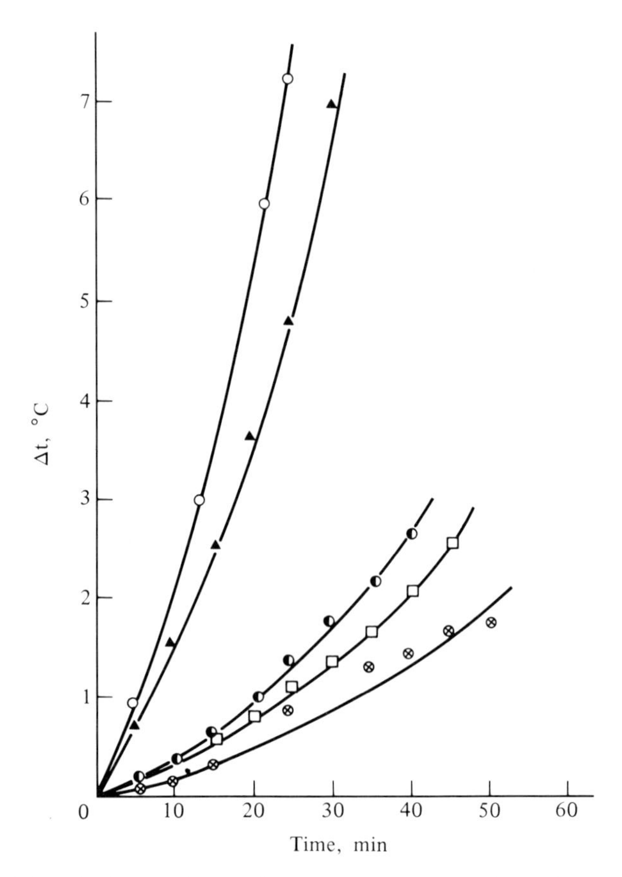

FIGURE 11. Polymerization reaction exotherms at various storage times of isoprene (over 99% purity) – *n*-heptane mixtures. ○ no storage; △ 48 hours; ◐ 96 hours; □ 144 hours; ⊗ 192 hours.

has been made in order to study the influence of the nature of the aluminium alkyl on the polymer characteristics. As is well known, organometallic compounds have a different activity depending on the nature of the organic groups attached to the aluminium atom, its degree of association in solution and on its alkylating ability which acts in the opposite direction to the degree of association.

Experiments with triisobutylaluminium and titanium tetrachloride catalyst system were performed in conditions similar to those used in polymerizations with the catalyst complex prepared from triethylaluminium and titanium tetrachloride. In order to compare the activity of the two catalyst systems, the variables determining the main features of *cis*-1,4-polyisoprene polymerization have been studied. The following variables have been investigated:

— Al/Ti molar ratio,
— catalyst concentration,
— nature of the solvent.

(*a*) *Influence of Al/Ti molar ratio.* To analyse the influence of the Al/Ti molar ratio on the microstructure, molecular weight, reaction rate and the physico-mechanical properties, the molar ratio was varied from 0.5 to 2, as in polymerizations with triethylaluminium and titanium tetrachloride. The other variables, i.e. a catalyst concentration of 1 g TIBA/100 g isoprene, an isoprene/*n*-heptane ratio of 1/4 (v/v), a reaction temperature of 32–35°C and a reaction time of 120 minutes, were kept constant. The results are given in Table 19.

The experimental data gathered in polymerizations carried out with the triisobutylaluminium catalyst system indicate that, depending on the Al/Ti molar ratio, polymers were obtained with properties similar to those prepared with the catalyst system containing triethylaluminium.

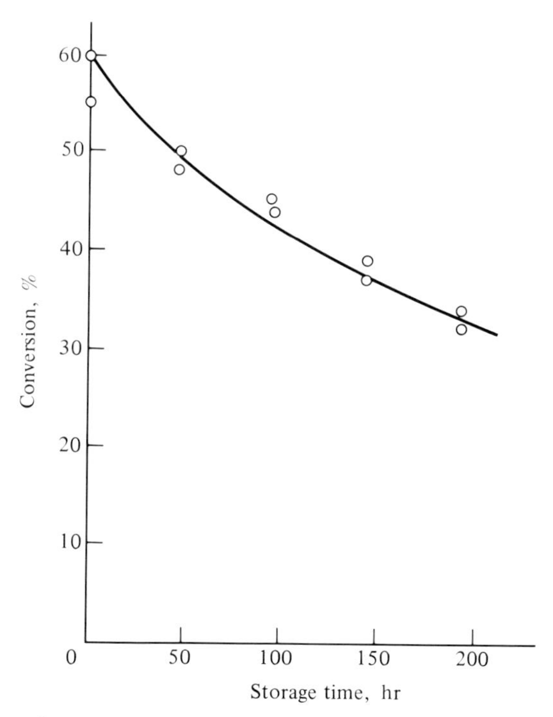

FIGURE 12. Effect of storage time of isoprene (over 99% purity)–*n*-heptane mixtures on conversion.

Table 19
Effect of Al/Ti molar ratio

No.	Al/Ti molar ratio	Conversion (%)	[η] (dl/g)	Properties of polyisoprene								Remarks
				Microstructure				Physico-mechanical properties				
				cis-1,4 (%)	*trans*-1,4 (%)	3,4 (%)	1,2 (%)	Tensile strength (kgf/cm^2)	Elongation at break (%)	Modulus at 300% elongation (kgf/cm^2)	Set at break (%)	
1	0.5	31	0.3	29	56	9	6	—	—	—	—	Granular polymer
2	0.5	34	0.25	27	57.5	9	6.5	—	—	—	—	
3	0.5	33	0.31	30	54.5	8.5	7	—	—	—	—	
4	1.0	78	4.0	92.5	0	5	2.5	268	700	14	9	
5	1.0	80	4.1	93	0	4	3	278	690	15	10	
6	1.0	81	3.9	94	0	4	2	265	710	15	8	
7	1.2	77	3.9	94	0	4	2	267	715	16	10	
8	1.2	75	3.7	92	0	5	3	258	720	15	10	
9	1.2	78	3.8	92.5	0	5	2.5	262	715	17	11	
10	1.5	35	0.8	92	0	4.5	3.5	—	—	—	—	Oily polymer
11	1.5	33	1.0	93	0	4	3	—	—	—	—	
12	1.5	36	0.9	92.5	0	4	3.5	—	—	—	—	
13	2.0	6	0.1	92.5	0	4.5	3	—	—	—	—	Oily polymer
14	2.0	4.5	0.12	93	0	4	3	—	—	—	—	
15	2.0	3	0.1	92	0	4.5	3.5	—	—	—	—	

Over 99% isoprene fraction.
1 g TIBA/100 g isoprene.

In polymerizations at an Al/Ti molar ratio of 0.5, polymers with a very low content of *cis*-1,4 addition product (up to 30%) and about 55% of *trans* 1,4-addition product were obtained. The intrinsic viscosity ranged from 0.25 to 0.31 dl/g and conversion was about 34%. The polymers had a granular nature with a crosslinked structure and contained 35% insoluble material.

At Al/Ti molar ratios of 1.5 and 2, the polymers had a high content of *cis* 1,4 addition product of about 93%. The polymer molecular weight was low in both cases and the intrinsic viscosities were 0.9 and 0.1 dl/g, respectively. Conversion reached 35% at an Al/Ti molar ratio of 1.5 and dropped to 5% at an Al/Ti molar ratio of 2. Liquid polymers of an oily nature were obtained at the above molar ratios.

In polymerizations effected with an Al/Ti molar ratio of 1, polymers with improved characteristics, *ca.* 94% *cis*-1,4 addition product, intrinsic viscosities between 3.9 and 4.1 dl/g and conversions of about 80% were obtained. The physico-mechanical properties had high values, namely a tensile strength of 270 kgf/cm^2 and an elongation at break of 700%. Polymers with similar properties were obtained at an Al/Ti molar ratio of 1.2. The microstructure, molecular weights and conversions obtained at various Al/Ti molar ratios are shown in Figures 13, 14 and 15, respectively.

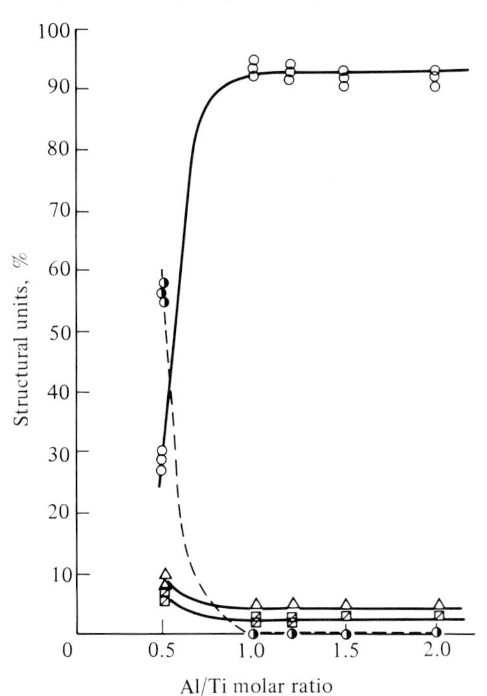

FIGURE 13. Effect of $TIBA/TiCl_4$ molar ratio on microstructure of polyisoprene (from isoprene fraction of over 99% purity). ○ *cis*-1,4; ◑ *trans*-1,4; △ 3,4; ◪ 1,2.

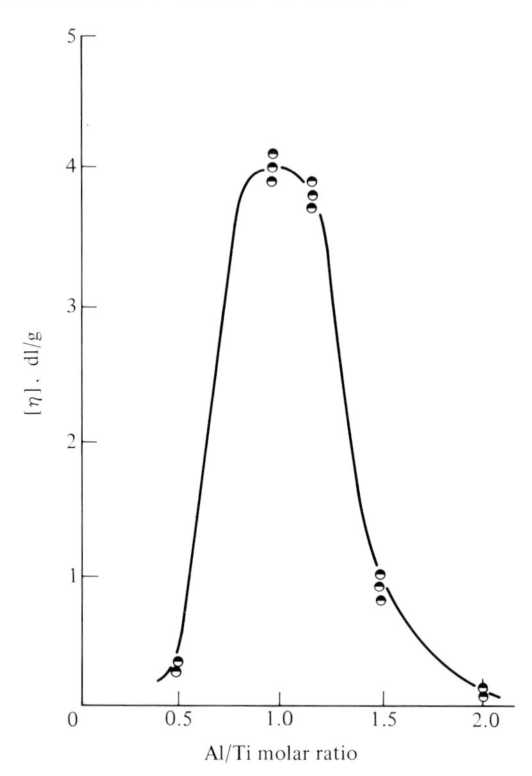

FIGURE 14. Effect of TIBA/$TiCl_4$ molar ratio on intrinsic viscosity of polyisoprene (from isoprene fraction of over 99% purity).

The diagrams show that the highest values for the yield of *cis*-1,4 addition product, molecular weights and conversions are obtained at an Al/Ti molar ratio of 1.

A comparison of exprimental data showing the influence of the Al/Ti molar ratio when triethylaluminium and triisobutylaluminium are used, leads to the conclusion that the polymers have optimum properties at the same Al/Ti molar ratio, although some differences are observed in the molecular weights and the yields. The catalyst complex containing triethylaluminium gives polymers with higher molecular weights, but the conversion is lower, while the catalyst complex containing triisobutylaluminium behaves in the opposite manner and results in the formation of polymers with a relatively lower molecular weight but with higher conversions. The differences are ascribed to the different activity of the catalyst systems investigated.

(*b*) *Influence of concentration of catalyst.* The influence of the concentration of triisobutylaluminium and titanium tetrachloride catalyst system on the properties of the polymer was also investigated. The catalyst concentration was varied from 0.5 to 2 g TIBA/100 g isoprene, as in the case of triethylaluminium, in order to compare the characteristics of the polymers prepared under similar reaction conditions, i.e. an Isoprene–*n*-heptane ratio of 1/4 (v/v), an Al/Ti molar ratio of 1, a reaction temperature of 32–35°C and a polymerization time of 120 minutes. The results with different catalyst concentrations are given in Table 20.

From the data shown in Table 20, we conclude that the catalyst concentration influences both the molecular weight and the reaction rate. Polymer stereoregularity is, however, not influenced by the catalyst concentration. The polymers obtained with various catalyst concentrations contain a high amount, more than 93 %, of *cis*-1,4 addition product.

The molecular weight is inversely proportional to the catalyst concentration. Experimental data indicate that molecular weight of the polymer decreases when the concentration of catalysts is increased from 1 to 2.5 g TIBA/100 g isoprene. At a catalyst concentration of 0.5 g TIBA/100 g isoprene, polymers with an intrinsic viscosity of 2.5 dl/g were obtained, lower

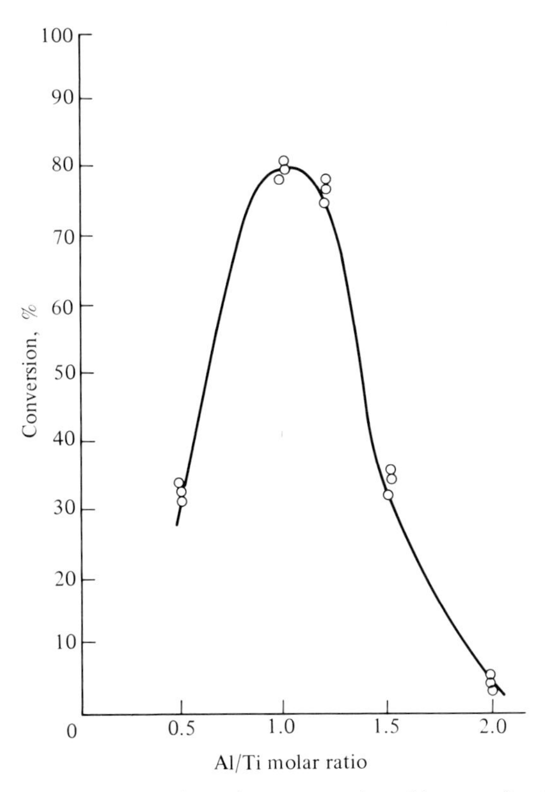

FIGURE 15. Effect of $TIBA/TiCl_4$ molar ratio on conversion of isoprene fraction of over 99 % purity.

Table 20

Effect of catalyst concentration

				Properties of polyisoprene							
				Microstructure				Physico-mechanical properties			
No.	g TIBA/100 g isoprene	Conversion (%)	$[\eta]$ (dl/g)	*cis*-1,4 (%)	*trans*-1,4 (%)	3,4 (%)	1,2 (%)	Tensile strength (kgf/cm^2)	Elongation at break (%)	Modulus at 300% elongation (kgf/cm^2)	Set at break (%)
1	0.5	50	2.5	91.5	0	5	3.5	220	755	19	13
2	0.5	52	2.4	92.5	0	5	2.5	235	740	20	12
3	0.5	54	2.6	93	0	4.5	2.5	230	745	20	12
4	1.0	78	4.0	92.5	0	5	2.5	268	700	14	9
5	1.0	80	4.1	93	0	4	3	278	690	15	10
6	1.0	81	3.9	94	0	4	2	265	710	16	8
7	1.5	83	3.5	92	0	5	3	260	690	15	9
8	1.5	85	3.5	94	0	4	2	252	720	18	11
9	1.5	84	3.6	92.5	0	4.5	3	259	710	17	10
10	2.0	88	2.8	93	0	4.5	2.5	224	740	21	11
11	2.0	90	2.6	92.5	0	5	2.5	219	770	20	12
12	2.0	89	2.7	91.5	0	5	3.5	232	735	20	12
13	2.5	92	2.2	94	0	4	2	191	785	21	12
14	2.5	94	2.0	92.5	0	5	2.5	187	800	22	13
15	2.5	93	2.1	93	0	4.5	2.5	183	810	20	15

Over 99% isoprene fraction.
Al/Ti molar ratio = 1.

than the value obtained at a catalyst concentration of 1 g TIBA/100 g isoprene where the intrinsic viscosity was 4 dl/g. At a catalyst concentration of 1.5 g TIBA/100 g isoprene, polymers were formed with somewhat lower molecular weights compared with those obtained with a concentration of 1 g TIBA/100 g isoprene. The corresponding intrinsic viscosity was 3.5 dl/g. The higher the catalyst concentration, the lower the molecular weight. Thus, at catalyst concentration of 2 and 2.5 g TIBA/100 g isoprene, the intrinsic viscosities were 2.6 and 2.1 dl/g, respectively.

The physico-mechanical properties also change with catalyst concentration, their values depending on the molecular weight. At a catalyst concentration of 1 g TIBA/100 g isoprene, the polymers had high molecular weights and hence improved physico-mechanical properties were obtained the tensile strength being 270 kgf/cm^2 and the elongation at break, 700 %.

The rate of the polymerization reaction increases proportionally with the concentration of the catalyst within the range of values studied. At a catalyst concentration of 0.5 g TIBA/100 g isoprene, conversions of about 50 % were obtained, while at a catalyst concentration of 1 g TIBA/100 g isoprene a marked increase of conversion was recorded, reaching 80 %. At higher catalyst concentrations, conversions increase by only 12 % in the large catalyst concentration interval ranging from 1 to 2.5 g TIBA/100 g isoprene.

The experimental data led to the conclusion that polymers with improved characteristics were prepared at a catalyst concentration of 1 g TIBA/100 g isoprene, in a similar manner to the catalyst system based on triethylaluminium.

The changes in molecular weight and reaction rate as a function of catalyst concentration are plotted in Figures 16 and 17.

A comparison of the molecular weights of the polymers prepared with the two catalyst systems at the optimum molar ratio of 1 leads to the conclusion that the physico-mechanical properties of polymers prepared with the catalyst system based on triisobutylaluminium would be unchanged, the values being close to those measured on polymers synthesized with the catalyst system based on triethylaluminium, although some lower molecular weights were obtained.

The reaction rate is influenced by the nature of the catalyst system. At the same catalyst concentration of 1 g TIBA/100 g isoprene conversion was about 20 % higher than with triethylaluminium.

(*c*) *Influence of nature of solvent.* The influence of the nature of the solvent on the polymerization of the isoprene of over 99 % purity with the catalyst system based on triisobutylaluminium and titanium tetrachloride and on properties of the polymer was also studied. Two non-polar solvents of saturated aliphatic hydrocarbon type, *n*-heptane and a hexane fraction, were used in the experiments. The latter solvent was tested in order to replace the single solvent, *n*-heptane, with a boiling point at normal pressure of

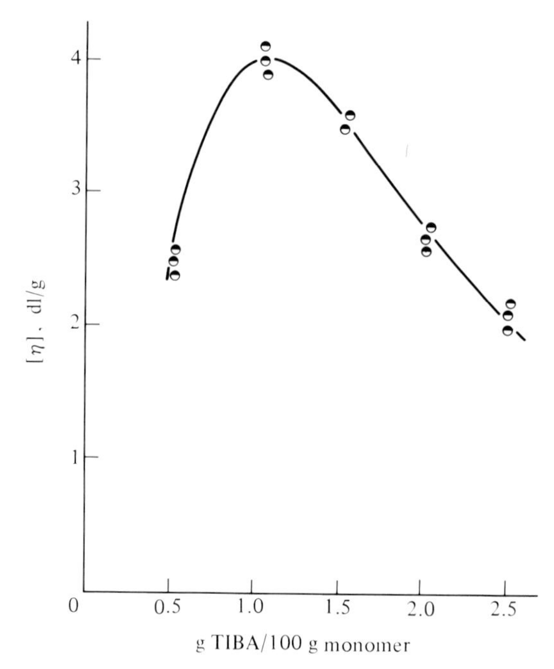

FIGURE 16. Effect of catalyst concentration on intrinsic viscosity of polyisoprene (from isoprene fraction of over 99 % purity).

98.2–98.4°C, with a hexane fraction with a wider boiling interval, i.e. 65 to 72°C. Therefore simultaneous polymerization experiments with both solvents of the same grade of purity were carried out. The runs were made under the optimum conditions previously established. The variables, i.e. an Al/Ti molar ratio of 1, a catalyst concentration of 1g TIBA/100g isoprene, a monomer/solvent ratio of 1/4 (v/v), a reaction temperature of 32–35°C and a polymerization reaction time of 120 minutes, were kept constant. The experimental results are shown in Table 21.

The above data indicate that *n*-heptane can be replaced with the hexane fraction since the properties of the polymers obtained by polymerization of isoprene of over 99 % purity in the latter solvent are not different from those of the polymers prepared in *n*-heptane. The polymers are highly stereoregular, containing more than 93 % of *cis*-1,4 addition product, the intrinsic viscosity is about 4 dl/g and the conversion 80 %. The physico-mechanical properties of the polymers show high values, i.e. tensile strengths of *ca.* 270 kgf/cm^2 and elongations at break of 700 %. The use of this solvent in polymerization is advantageous since the hexane fraction, having a lower boiling interval than *n*-heptane, can be more easily removed from the polymer solution.

Table 21
Effect of solvent

No.	Solvent	Con-version (%)	[η] (dl/g)	Properties of polyisoprene: Microstructure				Physico-mechanical properties			
				cis-1,4 (%)	*trans*-1,4 (%)	3,4 (%)	1,2 (%)	Tensile strength (kgf/cm^2)	Elong-ation at break (%)	Modulus at 300% elong-ation (kgf/cm^2)	Set at break (%)
1	*n*-hep-	78	4.0	92.5	0	5	2.5	268	700	14	9
2	tane	80	4.1	93	0	4	3	278	690	15	10
3		81	3.9	94	0	4	2	265	710	16	8
4		79	3.8	92.5	0	5	2.5	260	700	14	9
5	Hex-	80	4.0	91.5	0	5	3.5	259	710	17	10
6	ane	79	3.8	92.5	0	5	2.5	270	700	15	9
7	frac-	81	4.0	93	0	4.5	2.5	275	690	15	10
8	tion	79	3.9	94	0	4	2	260	700	15	9

Over 99% isoprene fraction
1 g TIBA/100 g isoprene.
Al/Ti molar ratio = 1.

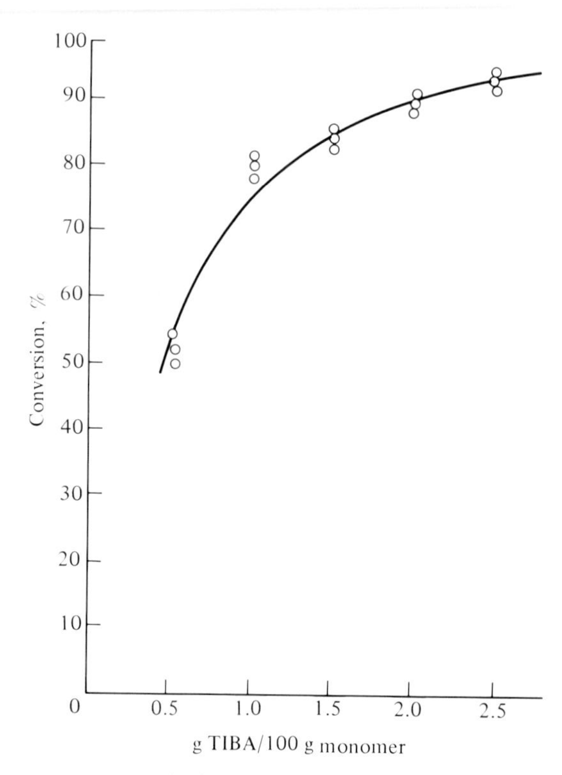

FIGURE 17. Effect of catalyst concentration on conversion of isoprene fraction of over 99% purity.

Discussion. In the polymerization of isoprene of over 99% purity, two catalyst systems prepared by complexing triethylaluminium and triisobutylaluminium, respectively, with titanium tetrachloride, were tested. Both catalyst systems were studied since the activity of the catalyst complex is dependent on the nature of the aluminium alkyl. Among the various aluminium alkyls, triethylaluminium and triisobutylaluminium were chosen since they show a convenient ratio between their degree of association in solution and their alkylating capacity, the differences in activity being attributed to the length of the alkyl chain. The influence of the Al/Ti molar ratio and the catalyst concentration on the properties of the polymer was investigated for both catalyst systems. The influence of the Al/Ti molar ratio was studied since it affects polymer microstructure, molecular weight and reaction rate. We tried to find the optimum value which produces a high degree of stereoregularity and a high molecular weight, both of which contribute to high values of physico-mechanical properties. With both catalyst systems, at an Al/Ti molar ratio of 1, polymers with a high content of *cis*-1,4 structural units, a high molecular weight and maximum reaction rate were obtained.

Investigation of the influence of the concentration of the catalyst on the molecular weight and the rate of reaction was made in order to establish the conditions of synthesis of polymers at convenient reaction rates with the highest molecular weights and improved physico-mechanical properties. Experimental data showed that with both catalyst systems a catalyst concentration of 1 g AlR_3/100 g isoprene leads to maximum values of molecular weight and to improved physico-mechanical properties. At the same catalyst concentration, the reaction rate varied according to the nature of the catalyst system, a faster reaction was observed with the triisobutyl-aluminium–titanium tetrachloride catalyst system owing to its prolonged activity with time. At an Al/Ti molar ratio of 1, the polymers had a high stereoregularity and the catalyst concentration had no effect on polymer microstructure.

The variation in time of the monomer purity and the influence of isoprene storage time on the polymerization process were also investigated. Experimental data indicated that storage produces a decrease of the reaction rate, emphasizing the need to purify isoprene before it is used in polymerization.

Stereospecific polymerization of isoprene in a hexane fraction, a solvent with a wide boiling range, was also studied in order to be able to replace *n*-heptane previously used. The characteristics of the polymers were similar to the ones prepared in *n*-heptane. Hence, it is possible to replace this solvent with the hexane fraction.

From experimental data gathered in the stereospecific polymerization of isoprene of over 99% purity, we noticed that the same correlations between Al/Ti molar ratio, catalyst concentration and polymer characteristics are valid for both catalyst systems leading to similar values of microstructure and molecular weight. The only observed difference between the two catalyst systems is a faster reaction rate in the case of triisobutylaluminium and titanium tetrachloride catalyst complex resulting in higher conversions, which is an advantageous feature of this catalyst.

The variation of reaction rates for both catalyst systems is plotted in Figure 18. These data emphasize the difference of the two catalyst systems.

B. Polymerization of the *ca.* 25% isoprene fraction

Investigation of the stereospecific polymerization of an isoprene fraction of about 25% concentration with catalyst systems of the $AlR_3 + TiCl_4$ type was made in order to synthesize *cis*-1,4-polyisoprene with characteristics similar to the polymer prepared by polymerization of isoprene of over 99% purity.

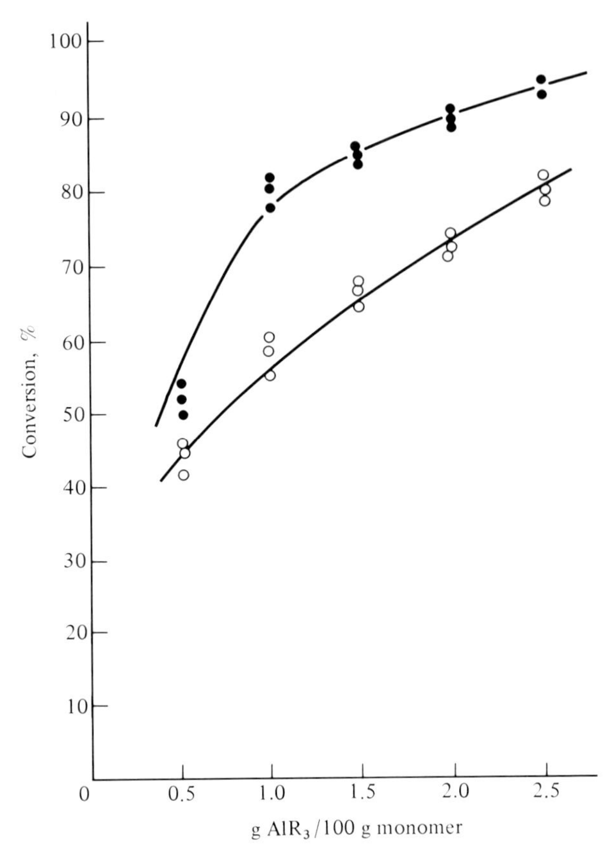

FIGURE 18. Effect of catalyst concentration on conversion of isoprene fraction of over 99% purity. ● TIBA + $TiCl_4$; ○ TEA + $TiCl_4$.

The main scope of this research was to replace with a monomer prepared by a much more simple procedure the highly concentrated monomer synthesized via the dehydrogenation of isopentane, followed by the successive purification operations necessary to reach this high degree of purity.

The isoprene fraction with a purity of about 25%, synthesized by the partial dehydrogenation of isopentane, contains *ca.* 63% of saturated hydrocarbons, mainly isopentane, and *ca.* 11% of olefins, mainly methylbutenes (Table 1 and Figure 2). In the polymerization of this isoprene, the addition of another solvent as a reaction medium is no longer necessary because of the high content of isopentane in the monomer fraction; the isopentane itself can be used as a solvent. Removal of the impurities interfering in the polymerization is concomitantly performed on the mixture of monomer and solvent. The use of the *ca.* 25% isoprene fraction in the polymerization process is advantageous since the monomer is prepared via a simplified procedure, it does not involve the separate preparation of a solvent and it reduces the number of purification steps.

In order to prepare a polyisoprene with a high content of *cis*-1,4 addition product by polymerization of the *ca.* 25 % isoprene fraction, we tried to use the same catalyst complex as for isoprene of over 99 % purity. As is well-known, Ziegler–Natta-type catalysts are able to catalyze the stereospecific polymerization of either dienes or olefins according to the ratio of the components in the catalyst. Thus, olefins polymerize at Al/Ti molar ratios ranging from 2 to 10, whereas the optimum value of the Al/Ti molar ratio for isoprene polymerization is unity. At this ratio, the transition metal is reduced to Ti(III) and catalyses the diene polymerization. In particular the catalyst complex formed from $AlR_3 + TiCl_4$ catalyses the polymerization of either olefin or diene to the Al/Ti molar ratio in the complex. Selective polymerization of isoprene in the presence of olefins can thus be achieved depending on the Al/Ti molar ratios used.

The synthesis of *cis*-1,4-polyisoprene by polymerization of an isoprene fraction of *ca.* 25 % purity with heterogeneous catalyst systems of the $AlR_3 + TiCl_4$ type has not been reported in the literature.

In polymerization experiments with the isoprene fraction of *ca.* 25 % purity we investigated the reaction conditions necessary to obtain polymers with elastomeric properties and a high degree of stereoregularity, similar to the polyisoprene produced by polymerization of isoprene of over 99 % purity. Two catalyst systems, i.e. triethylaluminium–titanium tetrachloride and triisobutylaluminium–titanium tetrachloride, were used. The influence of the Al/Ti molar ratio and the catalyst concentration was studied with both catalyst systems. The parameters were varied within the same ranges as in the polymerization studies performed with isoprene of over 99 % purity. The polymerization reaction was carried out at normal pressure under an inert gas (purified nitrogen), at a monomer/solvent ratio of 1/4 (v/v). The catalyst complex was prepared *in situ* and the reaction temperature was kept constant at 30–32°C, somewhat lower than in the polymerization of isoprene of 99 % purity. The reagents were purified before using in the polymerization reaction (Chapter I).

1. Polymerization of the isoprene fraction of ca. 25 % purity with triethylaluminium and titanium tetrachloride

In polymerization of the isoprene fraction of *ca.* 25 % purity with the triethylaluminium and titanium tetrachloride catalyst complex, the same variables as in the polymerization of isoprene of over 99 % purity were varied in order to study the properties of polymers which had been prepared under comparable conditions.

The following variables were studied:

— Al/Ti molar ratio,
— catalyst concentration.

(*a*) *Influence of Al/Ti molar ratio.* Since the polymer properties depend on the Al/Ti molar ratio, we tried to find the optimum molar ratio yielding polymers with high stereoregularity, molecular weight and degree of conversion. Experiments were carried out varying the Al/Ti molar ratio from 0.5 to 2. The catalyst concentration, 1 g TEA/100 g isoprene, the reaction temperature, 30–32°C, and the reaction time, 120 minutes, were kept constant. Characterization of the polymers was made by determining their microstructure, molecular weight and physico-mechanical properties. The results are given in Table 22. The IR absorption spectra of polyisoprene prepared at Al/Ti molar ratios of 0.5, 1 and 2 are shown in Figures 19a, b and c, respectively. The changes in microstructure, molecular weight and conversion as a function of Al/Ti molar ratio are shown in Figures 20, 21 and 22, respectively.

The experimental data recorded in the polymerization of the isoprene fraction of *ca.* 25 % purity with triethylaluminium + titanium tetrachloride catalyst system at various Al/Ti molar ratios, emphasize its influence on the properties of the polymer.

In polymerizations carried out at an Al/Ti molar ratio of 0.5, polymers with a very low content of *cis*-1,4 addition product (29 %) and with more than 50 % *trans*-1,4 product were obtained. At this ratio, the intrinsic viscosity of the polymers was very low, about 0.2 dl/g, and the degree of conversion did not exceed 27 %. The polymers had a crosslinked structure and a granular appearance; the content of insoluble material was about 30 %.

At an Al/Ti molar ratio of 1, polymers with a high stereoregularity were obtained. The content of *cis*-1,4 addition structural units was 94 %. A high intrinsic viscosity ranging from 4.5 to 4.9 dl/g and a conversion of about 60 % were obtained. The physico-mechanical properties also had high values: tensile strengths of about 270 kgf/cm^2 and elongations at break of 700 %. In polymerizations made at an Al/Ti molar ratio of 1.2, polymers with very similar properties to those prepared at an Al/Ti molar ratio of 1 were obtained.

The polymers synthesized at an Al/Ti molar ratio of 1.5 had the same content of *cis*-1,4-structural units, but the intrinsic viscosity was only 1.6 dl/g and the reaction rate was very slow; the final conversion reached only 25 %.

At an Al/Ti molar ratio of 2, the polymers had a lower intrinsic viscosity, only 0.25 dl/g, and conversion was only *ca.* 4 %. However, the content of *cis*-1,4 addition product remained high. The polymers were oily in nature and their physico-mechanical properties could not be measured.

Experiments made in order to establish the optimum Al/Ti molar ratio

Table 22
Effect of Al/Ti molar ratio

				Properties of polyisoprene								
				Microstructure				Physico-mechanical properties				
No.	Al/Ti molar ratio	Conversion (%)	[η] (dl/g)	*cis*-1,4 (%)	*trans*-1,4 (%)	3,4 (%)	1,2 (%)	Tensile strength (kgf/cm^2)	Elongation at break (%)	Modulus at 300% elongation (kgf/cm^2)	Set at break (%)	Remarks
1	0.5	22	0.02	28	57	9.5	5.5	—	—	—	—	Granular polymer
2	0.5	27	0.16	29	56	9	6	—	—	—	—	
3	0.5	25	0.2	27	58	8.5	6.5	—	—	—	—	
4	1.0	61	4.7	92.5	0	5	2.5	270	700	14	9	
5	1.0	58	4.5	94	0	4	2	265	710	15	9	
6	1.0	57	4.9	93	0	4	3	280	690	16	8	
7	1.2	55	4.3	94	0	4	2	265	710	17	8	
8	1.2	52	4.0	92	0	5	3	262	718	18	8	
9	1.2	54	4.1	93	0	4.5	2.5	260	715	18	8	
10	1.5	22	1.6	93	0	4	3	—	—	—	—	Oily polymer
11	1.5	25	1.8	93.5	0	4	2.5	—	—	—	—	
12	1.5	28	1.4	92	0	4.5	3.5	—	—	—	—	
13	2.0	5.2	0.25	92.5	0	5	2.5	—	—	—	—	Oily polymer
14	2.0	3.6	0.3	92	0	5	3	—	—	—	—	
15	2.0	4	0.2	94	0	4	2	—	—	—	—	

Ca. 25% isoprene.
1 g TEA/100 g isoprene.

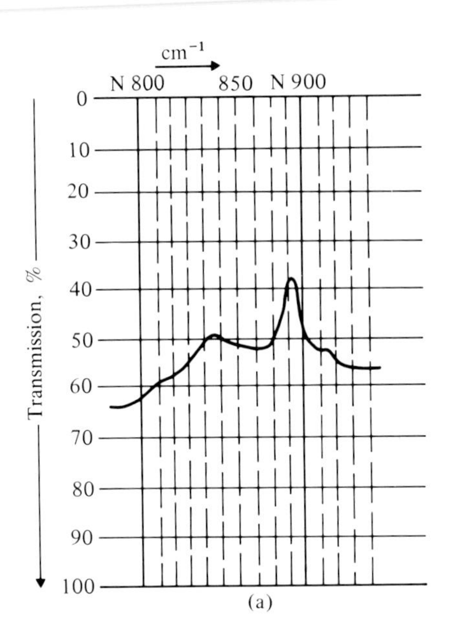

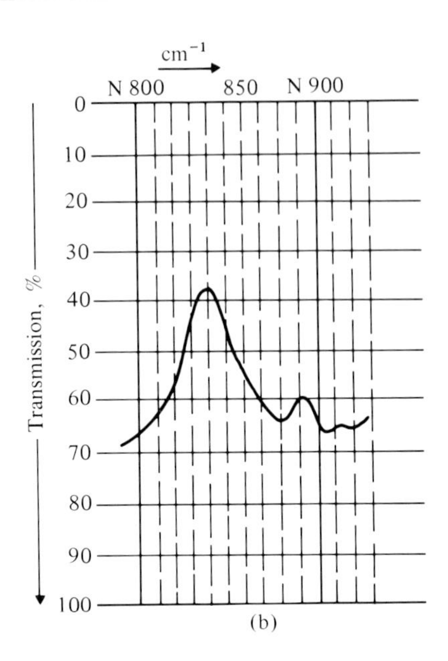

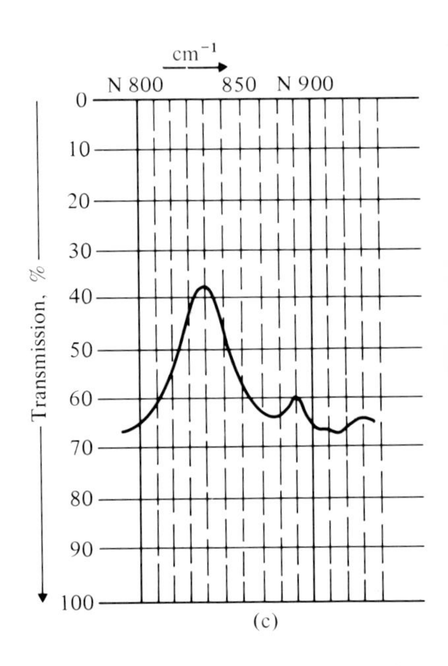

FIGURE 19. IR spectra of polyisoprene synthesized at various Al/Ti molar ratios from isoprene of *ca.* 25% purity. (a) Al/Ti molar ratio = 0.5. *cis*-1,4 = 27%; *trans*-1,4 = 58%; 3,4 = 8.5%; 1,2 = 6.5%. (b) Al/Ti molar ratio = 1. *cis*-1,4 = 93%; *trans*-1,4 = 0%; 3,4 = 4.5%; 1,2 = 2.5%. (c) Al/Ti molar ratio = 2. *cis*-1,4 = 92%; *trans*-1,4 = 0%; 3,4 = 5%; 1,2 = 3%.

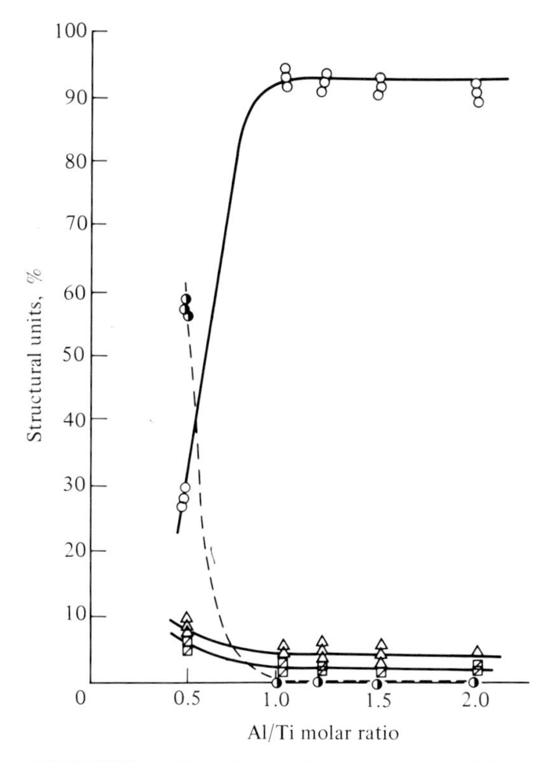

FIGURE 20. Effect of $TEA/TiCl_4$ molar ratio on microstructure of polyisoprene (from isoprene fraction of *ca.* 25% purity). ○ *cis*-1,4; ◑ *trans*-1,4; △ 3,4; ⊠ 1,2.

indicate that at a molar ratio of 1, polymers with the highest intrinsic viscosity and content of *cis*-1,4 structure were formed, as shown in Figures 20, 21 and 22.

A comparison between the characteristics of the polymers prepared by polymerization of the isoprene fraction of *ca.* 25% purity with the triethyl-aluminium–titanium tetrachloride catalyst system and the polymers synthesized from isoprene of over 99% purity shows that almost identical properties were obtained. The highest values resulted at an Al/Ti molar ratio of 1.

(*b*) *Influence of concentration of catalyst.* In polymerization of the isoprene fraction of *ca.* 25% purity with triethylaluminium–titanium tetrachloride catalyst system, we tried to find the catalyst concentration which results in the formation of polymers with improved characteristics whilst maintaining a convenient reaction rate. The catalyst concentration (based on monomer) was varied from 0.5 to 2.5g TEA/100g isoprene. The variables, i.e. an Al/Ti molar ratio of 1, a reaction temperature of 30–32°C, a polymerization reaction time of 120 minutes, were kept constant. The polymers were characterized by

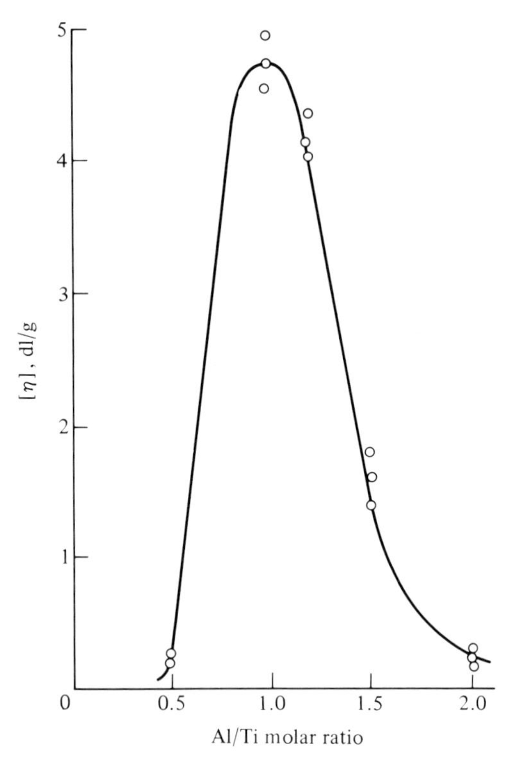

FIGURE 21. Effect of $TEA/TiCl_4$ molar ratio on intrinsic viscosity of polyisoprene (from isoprene fraction of *ca.* 25 % purity).

measuring the intrinsic viscosities and determining their microstructure and their physico-mechanical properties. The experimental results are given in Table 23. The variations in molecular weight and reaction rate with the concentration of the catalyst are plotted in Figures 23 and 24, respectively.

The data recorded in polymerizations made at various catalyst concentrations emphasize the influence of the catalyst concentration on certain polymer properties. The molecular weight and the reaction rate are affected by the catalyst concentration, but the polymer microstructure is, however, independent of the catalyst concentration. Polymers with a high content of *cis*-1,4 structural units (more than 93 %) were obtained within the whole range of catalyst concentrations examined. The polymers which were prepared at a catalyst concentration of 0.5 g TEA/100 g isoprene had an intrinsic viscosity of about 3 dl/g, an average conversion of 42 % was reached, and the following physico-mechanical properties were measured: tensile strength, 225 kgf/cm^2; elongation at break 730 %. At a catalyst concentration of 1 g TEA/100 g isoprene, polymers had a high intrinsic viscosity ranging from

4.5 to 4.9 dl/g, tensile strengths of 275 kgf/cm^2 and elongations at break of 700 %. Conversion varied between 57 and 61 %. At a catalyst concentration of 1.5 g TEA/100 g isoprene, an intrinsic viscosity of about 4 dl/g, an average conversion of 65 %, a tensile strength of 265 kgf/cm^2 and an elongation at break of 700 % were obtained. The use of catalyst concentrations of 2 and 2.5 g TEA/100 g isoprene led to polymers with low intrinsic viscosities of 3.2 and 2.5 dl/g, respectively, but conversions increased with increasing catalyst concentrations, reaching 72 and 80 %, respectively. The physico-mechanical properties had lower values with increasing catalyst concentrations.

A comparison of molecular weights, conversions and physico-mechanical properties measured at various catalyst concentrations leads to the conclusion that at concentrations of 1 and 1.5 g TEA/100 g isoprene, the polymers have the highest molecular weights and physico-mechanical properties, but conversions do not exceed 65 %.

As shown in Figure 23, the maximum molecular weight is reached at a catalyst concentration of 1 g TEA/100 g isoprene, as for polymers prepared from isoprene of over 99 % purity. Figure 24 shows the variation of the degree of conversion, which increases with the catalyst concentration.

The optimum catalyst concentration was found to be 1 g TEA/100 g isoprene. This concentration does not lead to a maximum degree of conversion, but produces the highest molecular weight, which in its turn provides the best physico-mechanical properties.

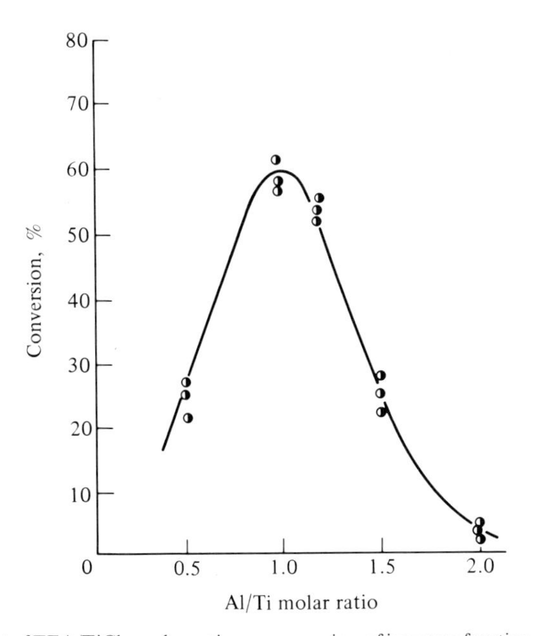

FIGURE 22. Effect of TEA/$TiCl_4$ molar ratio on conversion of isoprene fraction of *ca.* 25 % purity.

Table 23
Effect of catalyst concentration

				Properties of polyisoprene							
				Microstructure				Physico-mechanical properties			
No.	g TEA/100 g isoprene	Conversion (%)	$[\eta]$ (dl/g)	*cis*-1,4 (%)	*trans*-1,4 (%)	3,4 (%)	1,2 (%)	Tensile strength (kgf/cm^2)	Elongation at break (%)	Modulus at 300% elongation (kgf/cm^2)	Set at break (%)
1	0.5	39	3.1	93.5	0	4	2.5	235	725	19	14
2	0.5	44	2.7	92	0	5	3	220	735	20	12
3	0.5	42	2.8	92.5	0	5	2.5	225	748	18	14
4	1.0	60	4.7	92.5	0	5	2.5	270	700	14	8
5	1.0	58	4.5	94	0	4	2	265	710	15	9
6	1.0	57	4.9	93	0	4	3	280	690	16	8
7	1.5	63	4.0	93	0	4.5	2.5	270	700	16	9
8	1.5	67	3.8	92	0	4.5	3.5	260	710	15	9
9	1.5	65	3.9	94	0	4	2	262	705	15	8
10	2.0	74	3.2	92.5	0	5	2.5	240	730	18	14
11	2.0	70	3.3	93	0	4.5	2.5	245	725	19	14
12	2.0	72	3.0	92.5	0	5	2.5	230	745	20	12
13	2.5	78	2.7	92.5	0	5	2.5	220	740	20	12
14	2.5	79	2.5	94	0	4	2	215	775	19	12
15	2.5	82	2.3	93	0	4.5	2.5	195	785	21	12

Ca. 25% isoprene.
Al/Ti molar ratio = 1.

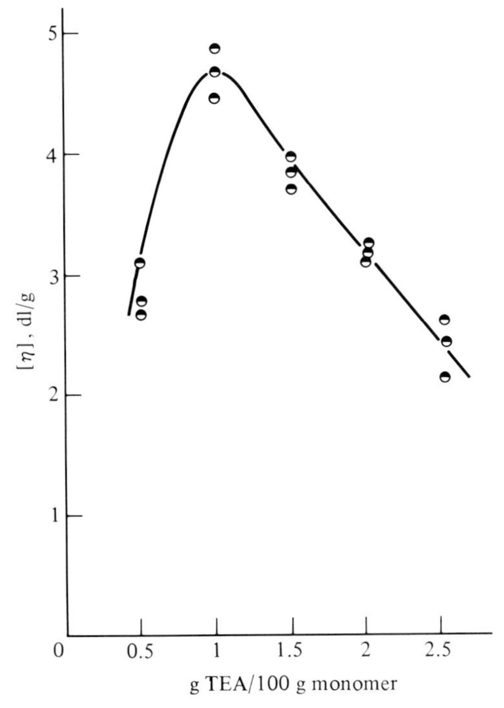

FIGURE 23. Effect of catalyst concentration on intrinsic viscosity of polyisoprene (from isoprene fraction of *ca.* 25% purity).

2. *Polymerization of the isoprene fraction of ca. 25% purity with triisobutylaluminium and titanium tetrachloride*

In polymerization experiments with the isoprene fraction of purity *ca.* 25% with the catalyst system based on triisobutylaluminium and titanium tetrachloride, the influence of the nature of the aluminium alkyl on the properties of the polymer was studied. Polymerizations with this catalyst system were carried out under similar conditions to those used with the isoprene of over 99% purity, in order to obtain comparative data. The same variables were investigated:

— Al/Ti molar ratio,
— catalyst concentration.

(*a*) *Influence of Al/Ti molar ratio.* In polymerizations of the isoprene fraction of purity *ca.* 25%, the influence of the Al/Ti molar ratio on polymer microstructure, molecular weight, physico-mechanical properties and reaction rate was investigated. The Al/Ti molar ratio was varied within the same range, i.e. 0.5 to 2, as for the isoprene of over 99% purity, while the

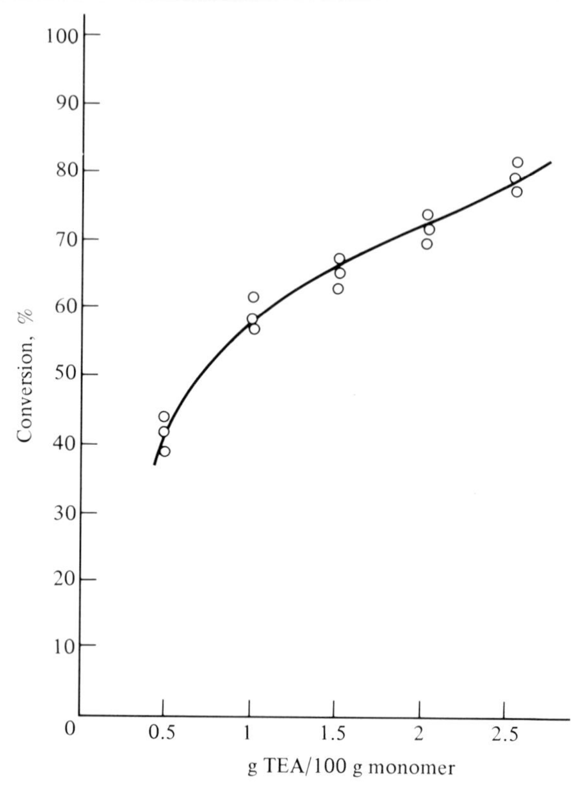

FIGURE 24. Effect of catalyst concentration on conversion of isoprene fraction of *ca.* 25 % purity.

catalyst concentration, 1 g TIBA/100 g isoprene, the reaction temperature, 30–32°C, and the reaction time, 120 minutes, were kept constant. The results are given in Table 24. The changes in microstructure, molecular weight and conversion as a function of the Al/Ti molar ratio are shown in Figures 25, 26 and 27, respectively.

From the experimental data we concluded that polymers with various properties depending on the Al/Ti molar ratio are obtained, as shown by the data listed in Table 24. Thus, at an Al/Ti molar ratio of 0.5, the polymers had a low content of *cis*-1,4 structural units (30 %), a very low intrinsic viscosity (0.2 dl/g) and an average conversion of only 32 %. The physico-mechanical properties could not be measured because the polymers were crosslinked and the content of insoluble material was 30 %.

At an Al/Ti molar ratio of 1, the highest values were obtained, namely 94 % *cis*-1,4 structure, an intrinsic viscosity of 4.2 dl/g, improved physico-mechanical properties and a conversion of 82 %. At an Al/Ti molar ratio of 1.2, high values of polymer properties were also obtained.

Table 24
Effect of Al/Ti molar ratio

No.	Al/Ti molar ratio	Conversion (%)	$[\eta]$ (dl/g)	Properties of polyisoprene: Microstructure: *cis*-1,4 (%)	*trans*-1,4 (%)	3,4 (%)	1,2 (%)	Physico-mechanical properties: Tensile strength (kgf/cm^2)	Elongation at break (%)	Modulus at 300% elongation (kgf/cm^2)	Set at break (%)	Remarks
1	0.5	31	0.15	30	55	9.5	5.5	—	—	—	—	Granular polymer
2	0.5	34	0.2	28	56.5	9.5	6	—	—	—	—	
3	0.5	30	0.18	31	53.5	9	6.5	—	—	—	—	
4	1.0	82	3.8	94	0	4	2	270	700	15	9	
5	1.0	79	4.2	92.5	0	5	2.5	278	690	15	10	
6	1.0	80	4.0	93	0	4	3	265	705	14	9	
7	1.2	78	3.9	92	0	5	3	265	710	16	10	
8	1.2	80	4.0	93	0	4	3	259	715	15	10	
9	1.2	77	3.7	94	0	4	2	263	705	17	11	
10	1.5	37	1.4	92	0	5	3	—	—	—	—	Oily polymer
11	1.5	35	1.3	92.5	0	4.5	3	—	—	—	—	
12	1.5	32	1.5	94	0	4	2	—	—	—	—	
13	2.0	8.2	0.2	93	0	4	3	—	—	—	—	Oily polymer
14	2.0	6	0.3	94	0	4	2	—	—	—	—	
15	2.0	7.5	0.25	92.5	0	4.5	3	—	—	—	—	

Ca. 25% isoprene.
1 g TIBA/100 g isoprene.

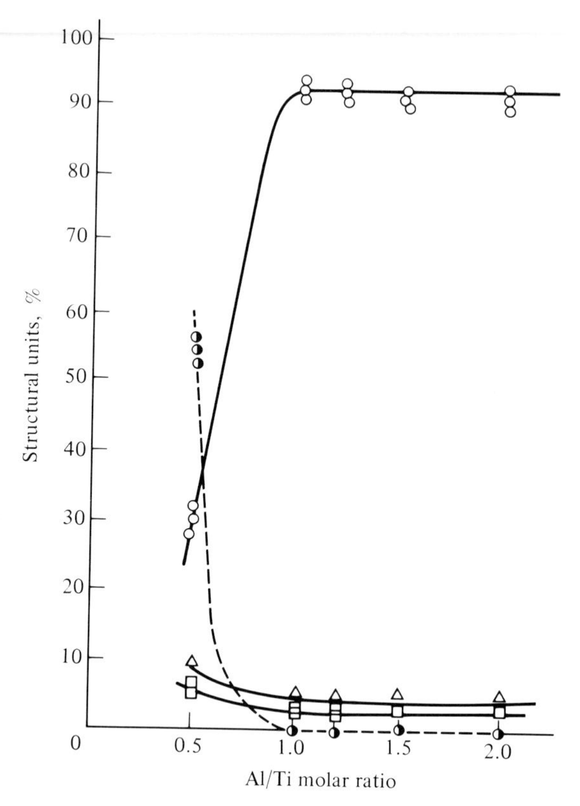

FIGURE 25. Effect of TIBA/$TiCl_4$ molar ratio on microstructure of polyisoprene (from isoprene fraction of *ca.* 25% purity). ○ *cis*-1,4; ◑ *trans*-1,4; △ 3,4; □ 1,2.

In polymerization experiments at Al/Ti molar ratios of 1.5 and 2, polymers were obtained with lower characteristics as compared with those recorded at Al/Ti molar ratios of 1 and 1.2. The molecular weights and conversions decrease with increasing Al/Ti molar ratio. The intrinsic viscosity was 1.4 dl/g at an Al/Ti molar ratio of 1.5, and 0.3 dl/g at an Al/Ti molar ratio of 2, while the degree of conversion was 35% and 6%, respectively. The polymer microstructure was not influenced by increase of the Al/Ti molar ratio within this range. The content of *cis*-1,4 structure was more than 92%. The polymers prepared under the above-mentioned conditions are oily products.

Optimum values for microstructure, molecular weight, conversion and physico-mechanical properties were obtained at an Al/Ti molar ratio of 1, as shown in Figures 25, 26 and 27.

The polymers yielded under such conditions have properties similar to those prepared from the isoprene of purity over 99%.

(*b*) *Influence of concentration of catalyst.* In the polymerization of the isoprene fraction of purity *ca.* 25% with the triisobutylaluminium and titanium tetrachloride catalyst system, the catalyst concentration was varied within the same range as in the polymerizations of isoprene of purity over 99 % in order to compare the influence of the catalyst concentration on the polymer properties in both cases.

The catalyst concentration was varied from 0.5 to 2.5 g TIBA/100 g isoprene, whereas the other variables were kept constant, i.e. an Al/Ti molar ratio of 1, a reaction temperature of 30–32°C, a polymerization reaction time of 120 minutes. The intrinsic viscosities, microstructure and the physico-mechanical properties of the polymers were determined. Experimental results are listed in Table 25. The variation of the molecular weight and reaction rate with catalyst concentration is plotted in Figures 28 and 29, respectively.

The experimental results show that the catalyst concentration influences both the molecular weight and the reaction rate. The microstructure of the polymer was not influenced within the range studied. Thus, in the concentration range from 0.5 to 2.5 g TIBA/100 g isoprene, the intrinsic viscosity reached a maximum value at a concentration of 1 g TIBA/100 g isoprene, further decreasing with an increase of catalyst concentration. The

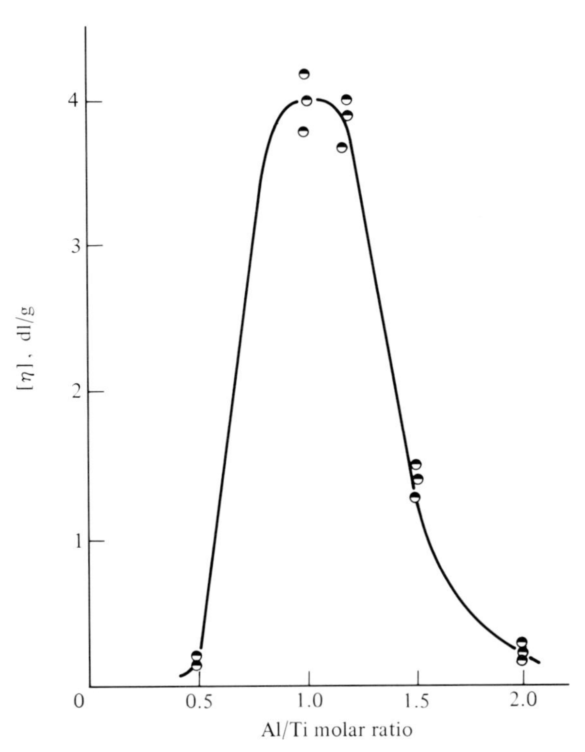

FIGURE 26. Effect of $TIBA/TiCl_4$ molar ratio on intrinsic viscosity of polyisoprene (from isoprene fraction of *ca.* 25 % purity).

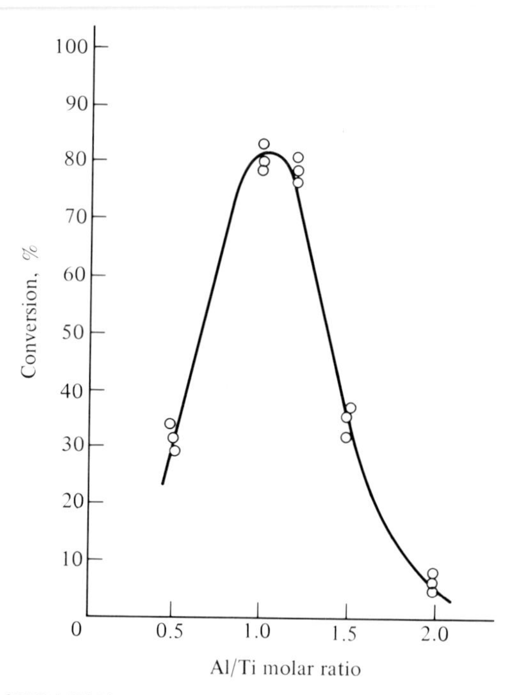

FIGURE 27. Effect of $TIBA/TiCl_4$ molar ratio on conversion of isoprene fraction of *ca.* 25 % purity.

reaction rate increased proportionally to the catalyst concentration. The polymers obtained at a catalyst concentration of 0.5 g TIBA/100 g isoprene had poor characteristics, i.e. an intrinsic viscosity not exceeding 2.7 dl/g, a conversion of 54 %, a tensile strength of 220 kgf/cm^2, and an elongation at break of 750 %. The highest molecular weight was obtained at a catalyst concentration of 1 g TIBA/100 g isoprene, with an intrinsic viscosity of 4.1 dl/g and improved physico-mechanical properties (a tensile strength of 270 kgf/cm^2 and an elongation at break of 700 %). The degree of conversion was also high, namely 82 %. Polymers with good characteristics were also obtained at a catalyst concentration of 1.5 g TIBA/100 g isoprene. At higher catalyst concentrations of 2 and 2.5 g TIBA/100 g isoprene, the polymers had inferior properties, which decreased with an increase of catalyst concentration. Thus, intrinsic viscosities of 2.7 and 2.1 dl/g, tensile strengths of 220 and 190 kgf/cm^2 and elongations at break of 740 and 800 %, respectively, were measured. The reaction rate was proportional to the catalyst concentration; a maximum conversion (94 %) was reached at a catalyst concentration of 2.5 g TIBA/100 g isoprene.

The content of *cis*-1,4 units still remained high (more than 93 %), within the whole range of catalyst concentrations studied since an Al/Ti molar ratio of 1 was used.

Table 25
Effect of catalyst concentration

				Properties of polyisoprene							
				Microstructure				Physico-mechanical properties			
No.	g TIBA/100 g isoprene	Conversion (%)	$[\eta]$ (dl/g)	*cis*-1,4 (%)	*trans*-1,4 (%)	3,4 (%)	1,2 (%)	Tensile strength (kgf/cm^2)	Elongation at break (%)	Modulus at 300% elongation (kgf/cm^2)	Set at break (%)
1	0.5	49	2.4	92	0	5	3	205	780	20	12
2	0.5	51	2.7	92.5	0	5	2.5	220	735	20	12
3	0.5	54	2.5	93	0	4	3	215	775	19	11
4	1.0	82	3.8	94	0	4	2	270	700	15	9
5	1.0	79	4.2	92.5	0	5	2.5	278	690	15	10
6	1.0	80	4	93	0	4	3	265	705	14	9
7	1.5	84	3.4	92.5	0	5	2.5	265	710	18	11
8	1.5	83	3.6	94	0	4	2	268	705	17	10
9	1.5	86	3.5	93	0	4	3	260	715	18	10
10	2.0	87	2.9	94	0	4	2	225	740	21	11
11	2.0	90	2.6	93	0	4.5	2.5	218	750	20	12
12	2.0	88	2.7	92.5	0	5	2.5	230	740	19	12
13	2.5	91	1.9	93	0	4	3	190	800	22	13
14	2.5	94	2.2	94	0	4	2	195	790	21	12
15	2.5	92	2.1	92	0	5	3	185	810	20	15

Ca. 25% isoprene.
Al/Ti molar ratio = 1.

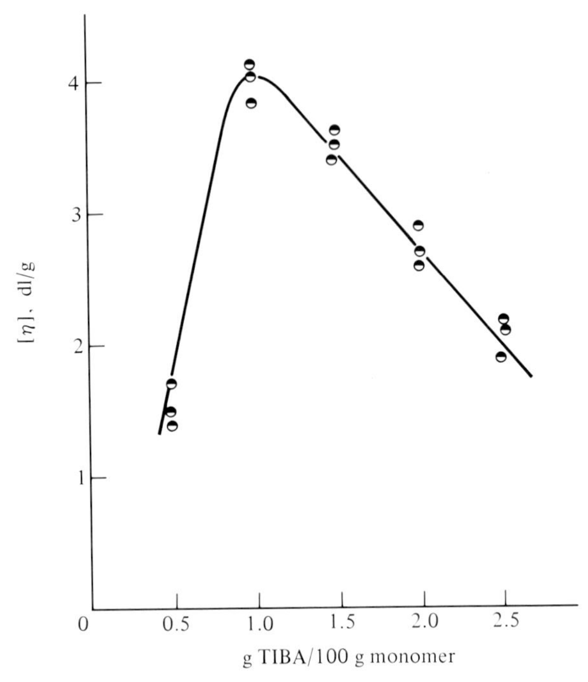

FIGURE 28. Effect of catalyst concentration on intrinsic viscosity of polyisoprene (from isoprene fraction of *ca.* 25% purity).

The variation of intrinsic viscosity and conversion with catalyst concentration, plotted in Figures 28 and 29, respectively, indicates that at a catalyst concentration of 1 g TIBA/100 g isoprene, a maximum molecular weight and a high conversion were obtained as in the case of the isoprene of purity over 99%.

Discussion. In the polymerization of the isoprene fraction of purity *ca.* 25% with the two catalyst systems, triethylaluminium or triisobutylaluminium and titanium tetrachloride, the influence of the Al/Ti molar ratio and of the concentration of the catalyst on the properties of the polymer was investigated. Experimental data indicate that high values of microstructure, molecular weight and conversion are obtained at an Al/Ti molar ratio of 1 and a catalyst concentration of 1 g AlR_3/100 g of isoprene for both catalyst systems. The physico-mechanical properties, which depend on the microstructure and the molecular weight of the polymer, reach the highest values in such conditions.

The activity of the catalyst systems used in polymerization of the isoprene fraction of purity *ca.* 25% is the same as in the polymerization of isoprene of purity over 99%. The catalyst complex based on triethylaluminium yields polymers with higher molecular weight but at lower conversions. On the other hand, the catalyst complex based on triisobutylaluminium appears to be more advantageous since faster reaction rates were recorded. The differences in molecular weight do not affect the physico-mechanical properties.

The data gathered in polymerization experiments of both isoprene fractions led to the conclusion that the polymers have similar characteristics, the highest values resulting at the same Al/Ti molar ratio and catalyst concentration.

The preparation of *cis*-1,4-polyisoprene from the isoprene fraction of purity *ca.* 25% is more advantageous compared with the polymerization of the isoprene of purity over 99%. As shown above, the fraction of lower isoprene content is prepared via a simplified procedure and contains a high amount of isopentane. The latter plays the role of a solvent and all purification operations prior to polymerization are carried out concomitantly. The polymerization reaction takes place under adiabatic conditions, the temperature not exceeding 32°C, owing to the low boiling temperatures of the components present in the reaction mixture. The heat evolved in the exothermic reaction is partially removed by the latent heat of vaporization of the components and cooling is unnecessary. On the other hand, owing to the low boiling temperatures of the components, their removal from the polymer solution is

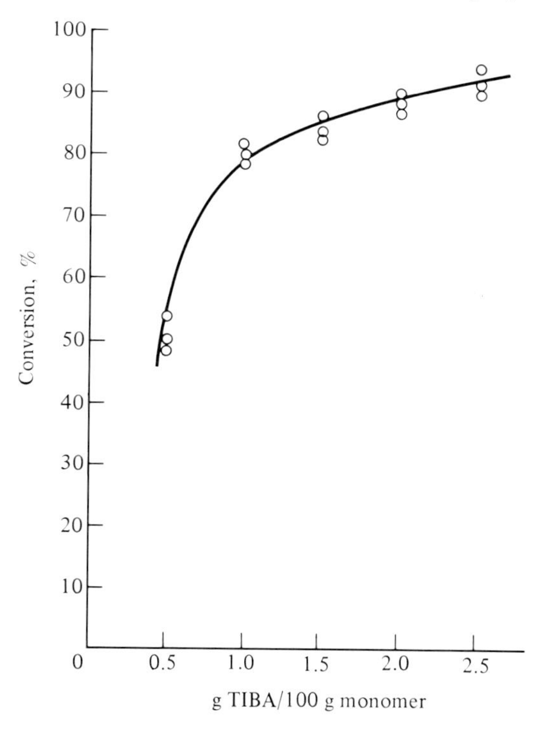

FIGURE 29. Effect of catalyst concentration on conversion of isoprene fraction of *ca.* 25% purity.

carried out at lower temperatures; the components separated from the polymer solutions, namely isopentane and olefins are introduced once more into the dehydrogenation process.

The polymerization of an isoprene fraction of *ca.* 25% purity with heterogeneous catalyst systems has not been described previously in the literature and a patent has therefore been granted [411]. Such studies have a theoretical interest as well, since under the reaction conditions used, a selective polymerization of isoprene in the presence of saturated and unsaturated (olefins) hydrocarbons is performed. As is well-known, heterogeneous catalysts of the $AlR_3 + TiCl_4$ type catalyse the stereospecific polymerization of either dienes or mono-olefins depending on the molar ratio of the catalyst components; this ratio ranges from 2 to 10 for olefins, while for isoprene the optimum Al/Ti molar ratio is 1 [317].

Based on this difference in Al/Ti molar ratio, which produces polymerization of either mono-olefins or dienes, it has been possible to control the polymerization reaction in order to achieve a selective polymerization of isoprene in the fraction of purity *ca.* 25%.

Among all homogeneous and heterogeneous catalyst systems effective in diene polymerization, the heterogeneous catalyst system based on $AlR_3 + TiCl_4$ was chosen since it produces a polyisoprene with a high content of *cis*-1,4 structural units. With this catalyst system, the polymerization reaction of the isoprene fraction of purity *ca.* 25% takes place at 30–32°C at normal pressure with a fast reaction rate.

Apart from a theoretical interest, the polymerization process of the isoprene fraction of purity *ca.* 25% with heterogeneous catalyst systems provides some economic advantages.

The monomer is prepared via a much more simplified procedure and purification is simultaneously effected for both the monomer and the solvent. The polymerization reaction occurs at normal pressure, the temperature remaining constant. Solvent removal from polymer solutions takes place at a low temperature and the solvent can be directly recirculated into the dehydrogenation process.

C. Some aspects of the kinetics of the stereospecific polymerization of isoprene

Kinetic studies of the stereospecific polymerization reaction meet with difficulties, especially with heterogeneous catalyst systems because it is virtually impossible to isolate and analyse the compounds formed in the reaction between the catalyst components, i.e. AlR_3 and $TiCl_4$. In the formation of a macromolecule the following steps must be considered:

— the initiation step, which produces the chain carrier of the system;

—the chain growing or propagation step. In kinetic treatments it is assumed that the rates of chain carrier formation and its consumption are equal. A quasi steady state is reached in this stage which maintains constant the overall polymerization rate;
—the termination reaction. In ionic coordination polymerizations the termination of the chain growth takes place by transfer reactions.

The difficulties met in kinetic treatments of ionic or ionic coordination polymerization reactions arise from the fact that the various steps of the polymerization process are incompletely separated and therefore only the overall polymerization rate has been usually studied. The polymerization rate, i.e. the monomer consumption rate, is given by the following equation [136]:

$$-\frac{d[M]}{dt} = k[M]^n[C]^m$$

where [M] is the monomer concentration, [C] the active catalyst concentration, $n = 1$ or 2 and $m \geqslant 1$.

A first order reaction ($n = 1$) with respect to monomer for a given catalyst composition (a constant Al/Ti molar ratio) was reported by several authors for the polymerization of isoprene with $AlR_3 + TiCl_4$ [317, 327].

The main factors influencing the kinetics of the polymerization reaction are the Al/Ti molar ratio, the nature and the purity of the reagents and the temperature of polymerization. Data on the effect of Al/Ti molar ratio, catalyst concentration, and the nature and purity of the reagents have been reported in previous works.

The influence of the temperature and the time of polymerization on the reaction rate and the molecular weight was studied in the polymerization of isoprene fractions of purities over 99% and *ca.* 25% with the triisobutyl-aluminium + titanium tetrachloride catalyst system.

1. *Influence of the temperature on reaction rate and molecular weight*

The influence of the temperature on the reaction rate and the molecular weight was studied in polymerization experiments effected at different temperatures. Thus, the polymerization of isoprene of purity over 99% was effected in the temperature interval from 10 to 55°C, while in the case of the isoprene fraction of purity *ca.* 25%, the temperature ranged from 10 to 35°C. Higher temperatures cannot be used at normal pressure owing to the low boiling temperatures of the components. In all such experiments, the following variables were kept constant: an Al/Ti molar ratio of 1, a catalyst concentration of 1 g TIBA/100 g isoprene, a monomer/solvent ratio of 1/4 (v/v), and a polymerization time of 120 minutes. The molecular weights and the microstructure of the polymers were determined. The results are listed in

Tables 26 and 27. The variation of the reaction rate, expressed by conversion, and of the molecular weight, expressed by intrinsic viscosity, obtained at the stated reaction temperatures is plotted in Figures 30 and 31, respectively.

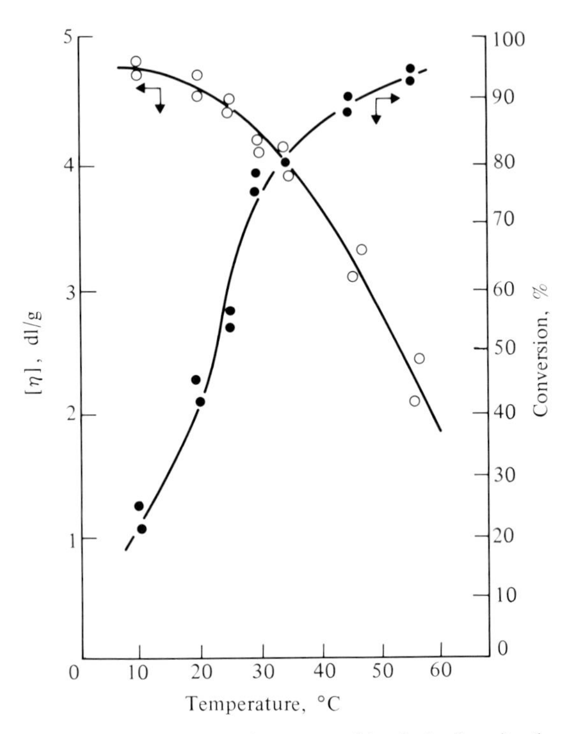

FIGURE 30. Effect of temperature on reaction rate and intrinsic viscosity (isoprene fraction of over 99% purity, TIBA). Al/Ti molar ratio = 1. 1 g TIBA/100 g isoprene.

The data recorded in the polymerization of isoprene of purity over 99% at various temperatures, shown in Table 26 and in Figure 30, emphasize the influence of temperature on both the polymerization rate and the molecular weight of the polymer. The reaction rate increases with the temperature, while the molecular weight decreases in the temperature range from 10 to 55°C. The intrinsic viscosity decreases from 4.8 to 2.1 dl/g, while conversion increases from 21 to 94%. A temperature of 30–35°C is the most advantageous since at this temperature the highest values for both the molecular weight and the conversion are obtained.

The same correlation between temperature, reaction rate and molecular weight was found in the polymerization of the isoprene fraction of purity *ca.* 25% as shown in Table 27 and Figure 31. The polymers prepared at temperatures ranging from 10 to 35°C had intrinsic viscosities varying in a

Table 26

Effect of temperature on reaction rate, molecular weight and microstructure (over 99% isoprene fraction)

No.	Reaction tempera- ture (%)	Con- version (%)	[η] (dl/g)	Microstructure *cis*-1,4 (%)	*trans*-1,4 (%)	3,4 (%)	1,2 (%)
1	10	25	4.8	93	0	4	3
2	10	21	4.7	92.5	0	5	2.5
3	20	45	4.7	92.5	0	5	2.5
4	20	42	4.5	93.5	0	4	2.5
5	25	54	4.4	94	0	4	2
6	25	56	4.5	92.5	0	4	3.5
7	30	75	4.1	93.5	0	4	2.5
8	30	78	4.2	94	0	4	2
9	35	80	4.1	93	0	4	3
10	35	81	3.9	94	0	4	2
11	45	88	3.3	93	0	4	3
12	45	90	3.1	92.5	0	5	2.5
13	55	94	2.4	93	0	4	3
14	55	93	2.1	92.5	0	4	3.5

Al/Ti molar ratio = 1.
1 g TIBA/100 g isoprene.

Table 27

Effect of temperature on reaction rate, molecular weight and microstructure (*ca.* 25% isoprene fraction)

No.	Reaction tempera- ture (%)	Con- version (%)	[η] (dl/g)	Microstructure *cis*-1,4 (%)	*trans*-1,4 (%)	3,4 (%)	1,2 (%)
1	10	20	4.8	93	0	4	3
2	10	18	4.9	93.5	0	4	2.5
3	15	32	4.7	94	0	4	2
4	15	30	4.8	92.5	0	5	2.5
5	20	42	4.6	93	0	4	3
6	20	44	4.7	92.5	0	4.5	3
7	25	57	4.3	94	0	4	2
8	25	59	4.2	93.5	0	4	2.5
9	30	79	4.1	94.5	0	3.5	2
10	30	80	4.0	93	0	4	3
11	35	82	3.8	93.5	0	4	2.5
12	35	84	3.9	94	0	4	2

Al/Ti molar ratio = 1.
1 g TIBA/100 g isoprene.

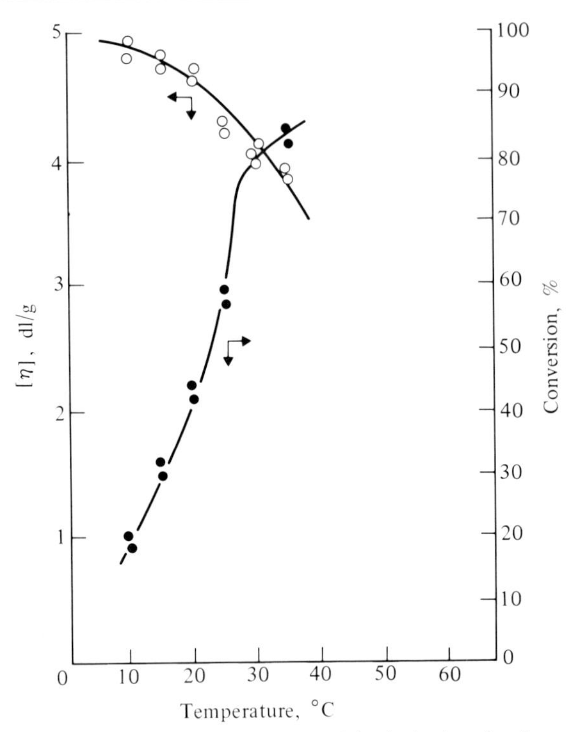

FIGURE 31. Effect of temperature on reaction rate and intrinsic viscosity (isoprene fraction of purity *ca.* 25 %, TIBA). Al/Ti molar ratio = 1. 1 g TIBA/100 g isoprene.

narrow range from 4.8 to 3.9 dl/g, while the degree of conversion took values between 20 and 84 %. In this case too, the temperature producing the highest values of both the molecular weight and conversion was 30–35°C. The polymerization reaction reached a steady state at this temperature owing to the boiling points of the components present in the reaction mixture.

Within the same temperature interval, high values for both the molecular weight and the reaction rate were reached, convenient for polymers prepared from both monomer fractions. At lower temperatures, 10°C, polymers with a higher intrinsic viscosity (4.8 dl/g) were formed; however, it was not advantageous to perform the reaction at this temperature because in such conditions the reaction was slow and low conversions of about 20 % were obtained.

2. Variation of the properties of polyisoprene with the polymerization time

In such experiments we tried to find the time necessary for the synthesis of *cis*-1,4-polyisoprene from both isoprene fractions. The following variables were kept constant in all the experiments: an Al/Ti molar ratio of 1, a catalyst concentration of 1 g TIBA/100 g isoprene, a monomer/solvent ratio of 1/4 (v/v), and a reaction temperature of 30°C. The polymerization reaction course was followed by taking samples after 30, 60, 120, 180, 240 and 300 minutes. Polymers were characterized by determination of their intrinsic viscosity and microstructure, samples being taken at the above time intervals. The results are listed in Tables 28 and 29 and the variations of the molecular weight and conversion for polymerization times are plotted in Figures 32 and 33, respectively.

From the data gathered in the polymerization of the isoprene of purity over 99 % we concluded that by increasing the polymerization time, an increase in conversion occurs. Thus, conversion reached the highest value of 94 % after a reaction time of 240 minutes, remaining constant after that up to 300 minutes. It was also proved that the reaction rate is faster in the first stage, conversion reaching 80 % after 120 minutes and after that it increases very slowly. This is explained by some authors by a decrease of monomer concentration on the solid catalyst surface at high conversions when the diffusion rate of the

Table 28

Variation of intrinsic viscosity, conversion and microstructure during polymerization (over 99 % isoprene fraction)

				Microstructure			
No.	Reaction time (min)	Conversion (%)	$[\eta]$ (dl/g)	*cis*-1,4 (%)	*trans*-1,4 (%)	3,4 (%)	1,2 (%)
1	30	32	3.5	93.5	0	4	2.5
2	30	35	3.4	93	0	4	3
3	30	33	3.5	92.5	0	4	3.5
4	60	50	3.95	94	0	4	2
5	60	52	4.0	92.5	0	5	2.5
6	60	55	3.98	93	0	4	3
7	120	78	4.0	93.5	0	4	2.5
8	120	80	4.1	93	0	4	3
9	120	81	3.9	94	0	4	2
10	180	85	3.92	93.5	0	4	2.5
11	180	86	3.95	94	0	4	2
12	180	84	4.0	93	0	4	3
13	240	94	3.98	92.5	0	5	2.5
14	240	92	4.0	94	0	4	2
15	240	93	4.05	93	0	4	3
16	300	92	4.0	93	0	4	3
17	300	92	4.08	92.5	0	4	3.5
18	300	94	3.98	94	0	4	2

Al/Ti molar ratio = 1.
1 g TIBA/100 g isoprene.

Table 29

Variation of intrinsic viscosity, conversion and microstructure during polymerization (*ca.* 25% isoprene fraction)

				Microstructure			
No.	Reaction time (min)	Con-version (%)	[η] (dl/g)	*cis*-1,4 (%)	*trans*-1,4 (%)	3,4 (%)	1,2 (%)
1	30	23	3.1	92.5	0	5	2.5
2	30	25	2.9	94	0	4	2
3	30	20	3.0	93	0	4	3
4	60	49	3.9	93.5	0	4	2.5
5	60	50	4.1	93	0	4	3
6	60	47	4.0	92.5	0	4.5	3
7	120	82	3.8	94	0	4	2
8	120	79	4.2	93.5	0	4	2.5
9	120	80	4.0	93	0	4	3
10	180	84	3.9	94	0	4	2
11	180	85	4.0	92.5	0	5	2.5
12	180	83	4.05	93	0	4	3
13	240	93	4.0	93.5	0	4	2.5
14	240	92	3.95	92.5	0	4.5	3
15	240	92	4.0	94	0	4	2
16	300	91	3.9	93	0	4	3
17	300	94	4.0	93.5	0	4	2.5
18	300	92	4.05	93	0	4	3

Al/Ti molar ratio = 1.
1 g TIBA/100 g isoprene.

monomer through the polymer layer becomes rate-determining [412].

The influence of polymerization time on molecular weight was also studied. Experimental results showed that molecular weight increased in the first reaction stage reaching the maximum value after 60 minutes and remained constant thereafter during the whole reaction time of 300 minutes.

Some papers also reported an initial increase of molecular weight which, after reaching the maximum value, remained constant. A sudden drop of viscosity at high conversions was attributed to certain degradation or branching reactions concomitant with the growing reaction. Such results have been recorded at higher temperatures (60°C) and longer reaction times (24 hours) [413].

Experimental data, shown in Table 29 and Figure 23, on the polymerization of the isoprene fraction of purity *ca.* 25%, indicate an increase of conversion with reaction time, reaching a maximum of 93% after 240 minutes, which remains constant thereafter.

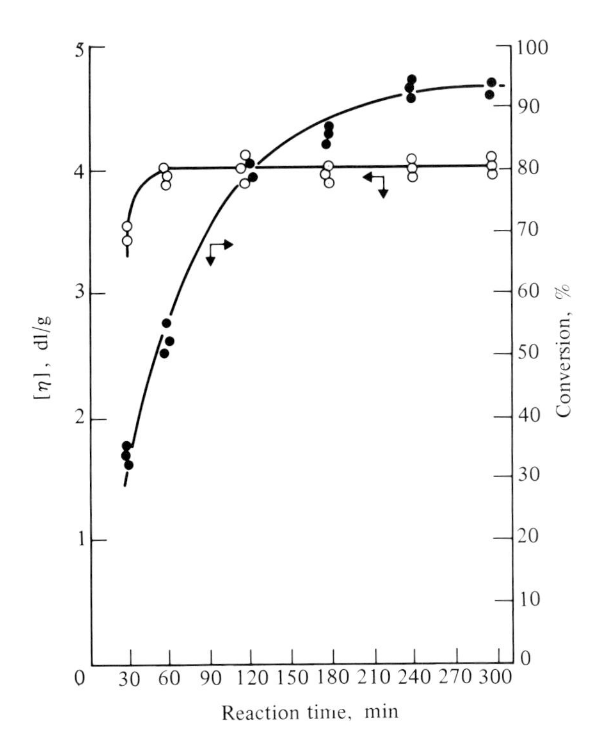

FIGURE 32. Variation of intrinsic viscosity and conversion during polymerization (isoprene fraction of over 99% purity, TIBA). Al/Ti molar ratio = 1. 1 g TIBA/100 g isoprene.

The molecular weight of the polymer also increases with the reaction time; it reaches a maximum after 60 minutes and then levels off.

Both the reaction rate and the molecular weight have lower values in the first polymerization stage of 30 minutes, compared with those observed in the polymerization of isoprene of over 99% purity, but they become equal after a reaction time of 60 minutes. This different behaviour during the first reaction stages was attributed to the nature of the solvent.

It was experimentally established that high molecular weights and conversions were obtained at a temperature of 30–35°C. It was also established that conversion reached 80% after a polymerization time of 120 minutes. For longer reaction times, conversion increased very slowly, which is not advantageous from an economic point of view.

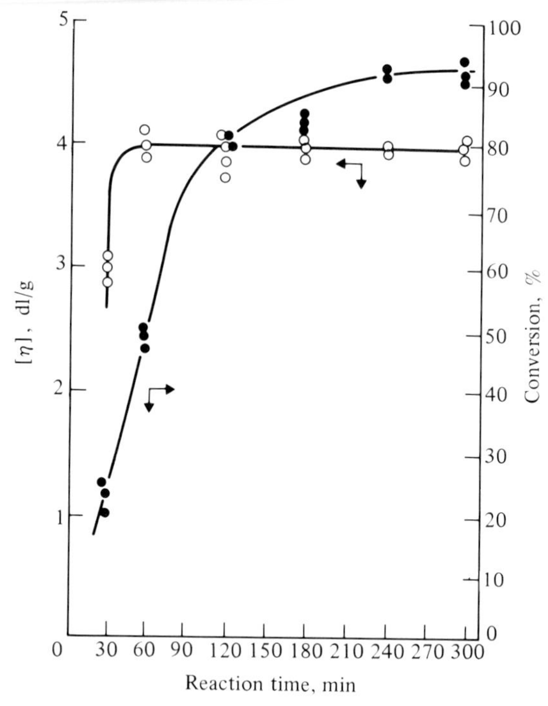

FIGURE 33. Variation of intrinsic viscosity and conversion during polymerization (isoprene fraction of *ca.* 25% purity, TIBA). Al/Ti molar ratio = 1. 1 g TIBA/100 g isoprene.

Polymer stereoregularity was not influenced by the reaction temperature and did not change during the polymerization process. The polymers prepared in such conditions had a predominant *cis*-1,4 structure in both cases.

Chapter III

Characterization of *cis*-1,4-polyisoprene

A. Physico-chemical characterization

The polymers prepared by polymerization of both isoprene fractions under the same reaction conditions showed very similar properties with regard to their microstructure and molecular weight as proved by analytical data. However, since monomer concentration and the content of saturated and unsaturated hydrocarbons were different in the two isoprene fractions, we considered it necessary to investigate also the IR absorption spectrum, the degree of unsaturation, the crystallinity and the molecular weight distribution. Measurements were made on polymers synthesized under the following conditions: an Al/Ti molar ratio of 1, a catalyst concentration of 1 g TIBA/100 g iosprene, a reaction temperature of 32°C, and a polymerization time of 120 minutes.

1. *Investigation of the microstructure of* cis-*1,4-polyisoprene*

The previously discussed IR spectra of the synthesized polymers were recorded with an UR 10 Zeiss–Jena IR absorption spectrophotometer in the 800 to 935 cm^{-1} range of the spectrum. The 838 cm^{-1} band is characteristic of a *cis*-1,4-structure, while the 842 cm^{-1} band is assigned to a *trans*-1,4 structure. At 890 cm^{-1} a vibration, assigned to a 3,4-addition structure, occurs and the peak at 910 cm^{-1} corresponds to a 1,2-structure [414–419]. The entire IR spectral range, from 700 to 4000 cm^{-1}, was recorded for a complete characterization of the polymers. The spectrum of the polymer prepared from isoprene of over 99 % purity is shown in Figure 34, while the one resulting from the isoprene fraction of *ca.* 25 % purity is given in Figure 35. The IR spectrum of a standard reference *cis*-1,4-polyisoprene (I) (also prepared with a Ziegler–Natta type catalyst complex), recorded under the same conditions, is shown in Figure 36.

A comparative study of the three rubber samples shows the presence of the absorption band assigned to the *cis*-1,4-structure at 838 cm^{-1}. The other bands within the 700–4000 cm^{-1} spectral range are virtually identical. The content of *cis*-1,4-structural units of all the polymers studied was identical.

2. *Determination of the degree of unsaturation of* cis-*1,4-polyisoprene*

The degree of unsaturation of samples of polymers prepared by the polymerization of the two specimens of isoprene was determined by chemical methods. Polymers prepared by diene polymerization have a high degree of

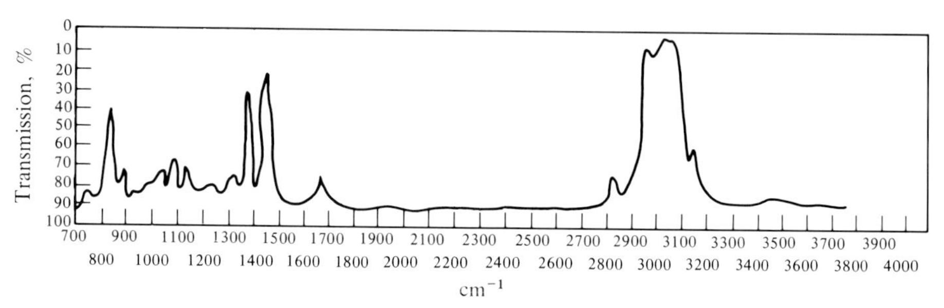

FIGURE 34. IR spectrum of polyisoprene obtained from isoprene fraction of over 99% purity.

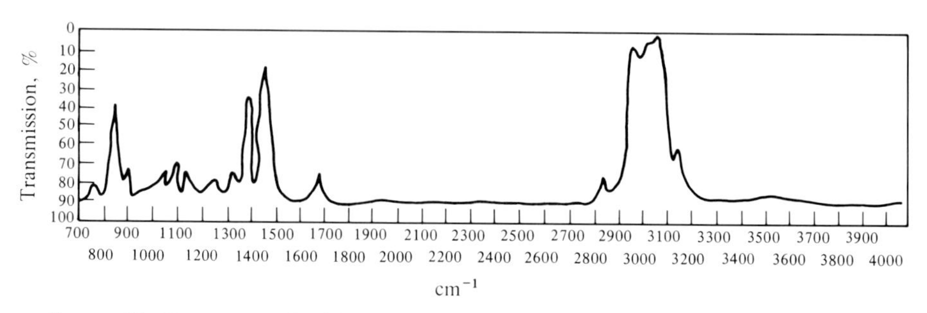

FIGURE 35. IR spectrum of polyisoprene obtained from isoprene fraction of *ca.* 25% purity.

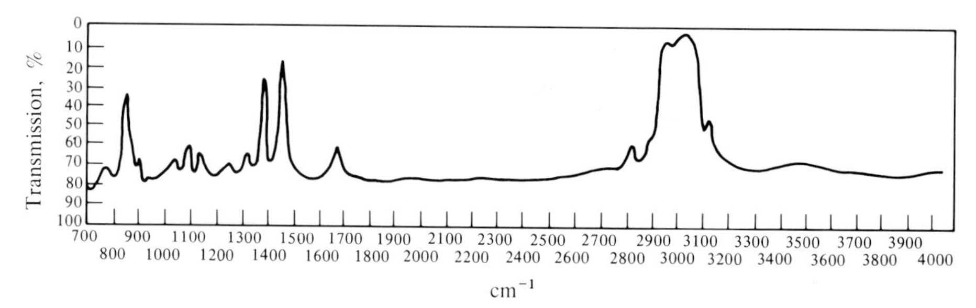

FIGURE 36. IR spectrum of standard *cis*-1,4-polyisoprene (I).

unsaturation owing to the double bonds present in the macromolecular backbone. In *cis*-1,4-polyisoprene, the double bonds occur between the C2 and C3 carbon atoms of the monomeric unit. The distance between two double bonds in the macromolecular backbone involves a sequence of four carbon atoms.

$$nCH_2{=}\underset{\displaystyle CH_3}{\underset{|}{C}}{-}CH{=}CH_2 \rightarrow$$

$$\sim CH_2{-}\underset{\displaystyle CH_3}{\underset{|}{C}}{=}CH{-}CH_2{-}CH_2{-}\underset{\displaystyle CH_3}{\underset{|}{C}}{=}CH{-}CH_2{-}CH_2{-}\underset{\displaystyle CH_3}{\underset{|}{C}}{=}CH{-}CH_2\sim$$

If a copolymer was formed between isoprene and an olefin present in the mixture of the isoprene fraction of purity *ca.* 25 %, e.g. 2-methyl-1-butene, the macromolecule would have the following chemical formula:

$$CH_2{=}\underset{\displaystyle CH_3}{\underset{|}{C}}{-}CH{=}CH_2 + CH_2{=}\underset{\displaystyle CH_3}{\underset{|}{C}}{-}CH_2{-}CH_3 \rightarrow$$

$$\sim CH_2{-}\underset{\displaystyle CH_3}{\underset{|}{C}}{=}CH{-}CH_2{-}CH_2{-}\overset{\displaystyle CH_3}{\overset{|}{\underset{\displaystyle \overset{\displaystyle CH_2}{\underset{|}{}}\atop CH_3}{\underset{|}{C}}}}{-}CH_2{-}\underset{\displaystyle CH_3}{\underset{|}{C}}{=}CH{-}CH_2\sim$$

In the copolymer the double bonds would be separated by distances of six carbon atoms, i.e. they would have a lower degree of unsaturation that the *cis*-1,4-polyisoprene macromolecules, where the double bonds are separated by only four carbom atoms.

The degree of unsaturation of polymers prepared from both isoprene fractions was measured by chemical methods. The double bonds were determined with iodine monochloride in glacial acetic acid (Kemp–Mueller's method [420]). The method is based on the addition of halogen to the double bonds present in the macromolecule. The same method was used to determine the degree of unsaturation of a standard *cis*-1,4-polyisoprene sample (I). The results are listed in Table 30.

Experimental data indicate a high degree of unsaturation, virtually equal for all the samples analysed. The above results prove that in polymerization of the isoprene fraction of purity *ca.* 25 %, copolymers with other unsaturated hydrocarbons present in the reaction mixture were not formed.

Table 30
Degree of unsaturation of *cis*-1,4-polyisoprene

Sample no.	Polyisoprene type	Unsaturation (%)
1	Polyisoprene from over 99% isoprene fraction	98
2		98.3
3		97.9
4	Polyisoprene from *ca.* 25% isoprene fraction	98.5
5		98
6		97.8
7	Standard *cis*-1,4-polyisoprene (I)	98

3. Determination of the degree of crystallinity of cis-*1,4-polyisoprene*

The elastomeric properties of polymers depend on the degree of crystallinity which is indicative of the arrangement of the macromolecular backbone chains. As is well-known, polymers can be classified into two groups: amorphous and partially crystalline. The amorphous state corresponds to a disordered arrangement of the macromolcules, while the crystalline state necessitates a high degree of order of the macromolecular chains [421]. Amorphous and crystalline polymers are easily differentiated by X-ray diffraction analysis. As a result of interaction of a strictly parallel X-ray beam ($\lambda = 1$–2 Å) with the electronic clouds of a molecule, a secondary emission of X-rays of the same wavelength occurs. This leads to the occurrence of maxima and minima in certain directions as the result of interference. In this way, the pattern of the atoms in the irradiated substance is made evident through the diffraction of the X-rays.

Under normal conditions, rubbers are amorphous, but in certain given conditions, by stretching and cooling, the chains tend to take an ordered linear form which corresponds to a crystalline state. Figure 37 shows the X-ray diagram of an amorphous rubber specimen and of a crystalline rubber, taken from ref. [422].

Diffraction patterns of amorphous and crystalline rubbers are different. An amorphous rubber gives a characteristic diffraction pattern with a diffuse cone around the central incident beam, while a crystalline rubber shows diffraction lines characteristic of the particular type of structure.

Determinations of crystallinity on unvulcanized rubber samples prepared from both the isoprene fractions were made with a TUR-M61 X-ray diffractometer. The samples were pressed into pellets (15 mm diam., 1.5 mm thick) and heated to 70°C to achieve homogeneity. A recording of the diffraction patterns was made photographically in a flat chamber and the presence of the crystalline phase was determined from the X-ray diagram. The

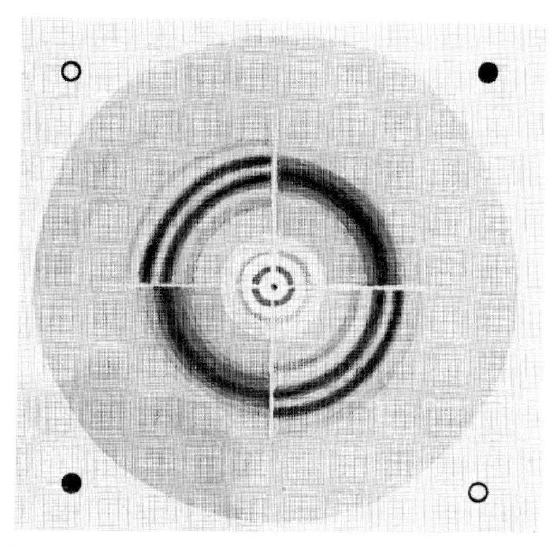

FIGURE 37. X-ray diffraction pattern of a sample of amorphous rubber ● and of a sample of crystalline rubber ○ [422].

degree of crystallinity was determined in this way on unvulcanized standard samples (types I and II were *cis*-1,4-polyisoprene prepared with Ziegler–Natta-type catalysts, type III was *cis*-1,4-polyisoprene prepared with butyllithium) and *cis*-1,4-polyisoprene synthesized in the present work. The crystallinity of natural rubber was also determined. All the X-ray diagrams are shown in Figure 38.

By examining the X-ray diagrams we concluded that all the rubber samples studied are identical, with a completely diffuse cone around the central incident beam, indicating the amorphous structure which is characteristic of elastomers.

4. Molecular weight distribution of cis-*1,4-polyisoprene*

The molecular weight of a polymer is determined by the molecular weight of the monomer and the number of monomeric units within the chain. The length of the molecules formed in polymerization reactions is distributed according to a probability function which depends on the reaction mechanism and the reaction kinetics. The molecular weight of the polymer is expressed as an average value since the number of monomeric units within the backbone chains is different from one macromolecule to another. The properties of a polymer are highly dependent on the distribution of the molecular weight, and it is therefore obviously necessary to know the molecular weight distribution of the polymer as well as the average molecular weight. As a guide to the distribution of polymer chain lengths, the ratio $\bar{M}_v/\bar{M}_n$ (the heterogeneity

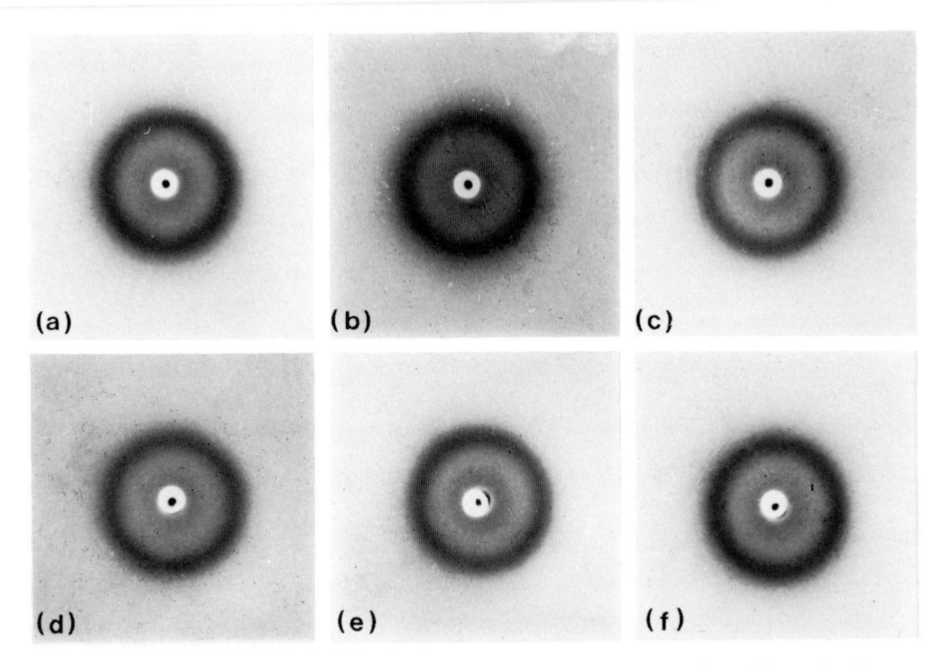

FIGURE 38. X-ray diffraction patterns of natural and syhthetic rubber samples. (a) Natural rubber. (b) Polyisoprene from isoprene of over 99% concentration. (c) Polyisoprene from isoprene of *ca.* 25% concentration; (d) *cis*-1,4-polyisoprene type I. (e) *cis*-1,4-polyisoprene type II; (f) *cis*-1,4-polyisoprene type III.

index) is frequently used. The larger the $\bar{M}_v/\bar{M}_n$ ratio, the wider the polydispersity. $\bar{M}_v$ is the viscosity-average molecular weight and $\bar{M}_n$, the number-average molecular weight.

The molecular weight distribution of *cis*-1,4-polyisoprene has scarcely been investigated. It is only reported that polymers of various degrees of polydispersity are prepared as a function of the nature of the catalyst system. Thus, polyisoprene prepared by polymerization with butyllithium has a narrow distribution of molecular weight, similar to that of a monodisperse polymer [423–425].

The polymers prepared in the presence of heterogeneous catalyst systems of the $AlR_3 + TiCl_4$ type have a wide molecular weight distribution, probably owing to the formation of different types of active catalyst centres [423, 426, 427].

The determination of the molecular weight distribution on polymers prepared from the two isoprene fractions was performed by fractional precipitation techniques using toluene as the solvent and methyl alcohol as the non-solvent [428].

The intrinsic viscosities of each polymer fraction were measured at 25°C in toluene solution with an Ubbelohde viscometer. It must be emphasised that, although polyisoprene is somewhat rapidly degradated in solution, this process could be avoided by using a procedure which provides fractionation in about 24 hours. This fact was verified by calculating the average of the intrinsic viscosities of the fractions, which was found equal to the intrinsic viscosity of the unfractionated polyisoprene. Thus, the measured viscosity of polyisoprene prepared from the isoprene of over 99 % purity was 4.1 dl/g, while the calculated value was 3.93 dl/g. In the case of polyisoprene prepared from the isoprene fraction of purity *ca.* 25 %, the measured viscosity was 4.0 dl/g and the calculated value 3.88 dl/g. Thus, based on the experimental data, integral and differential distribution curves were plotted by using the intrinsic viscosity scale. The integral and differential distribution curves for *cis*-1,4-polyisoprene prepared from isoprene of over 99 % purity are shown in Figure 39, while those of the polyisoprene prepared from the isoprene fraction of purity *ca.* 25 % are given in Figure 40.

The curves indicate a similarly wide molecular weight distribution in both polymer samples, in agreement with other data on the molecular weight distribution of polymers prepared with complex catalysts of the $AlR_3 + TiCl_4$ type.

Discussion. The properties of elastomeric *cis*-1,4-polyisoprene have been established by physico-chemical determinations. Thus, from the IR spectra we concluded that the polymers have a higher content of *cis*-1,4-addition structures. A high degree of unsaturation indicated the presence of double bonds in high proportion, which permitted a high degree of vulcanization to be

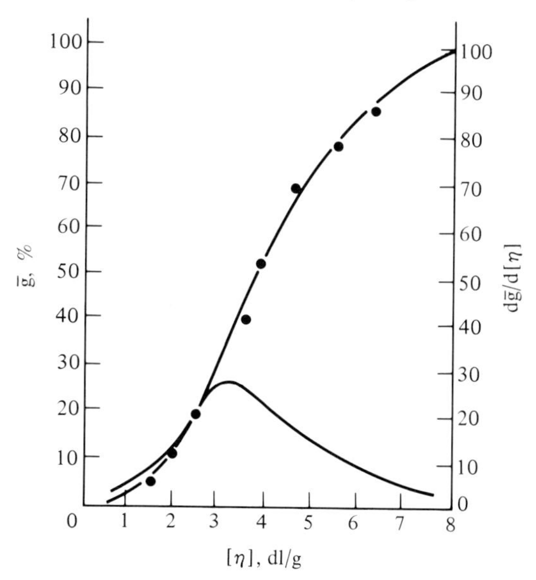

FIGURE 39. Integral and differential molecular weight distribution curves of *cis*-1,4-polyisoprene obtained from >99 % isoprene.

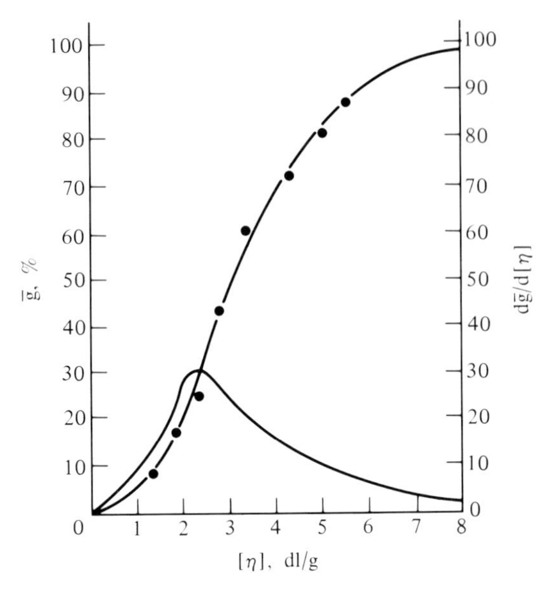

FIGURE 40. Integral and differential molecular weight distribution curves of *cis*-1,4-polyisoprene obtained from ~25% isoprene.

achieved and hence improved the elastic properties of the rubber. The arrangement of the macromolecular chains was indicated by X-ray diffraction. The diffraction patterns showed a diffuse cone around the incident central beam characteristic of the amorphous state, indicating an amorphous structure in the polymers studied which was typical of elastomers. The wide molecular weight distribution is typical of polymers prepared with heterogeneous catalyst systems.

B. Physico-mechanical characterization

The physico-mechanical properties of *cis*-1,4-polyisoprene depend on the physico-chemical properties of the polymer, on the recipe used in processing, and on the vulcanization conditions. In order to maintain the physico-chemical properties of the polymer obtained in the polymerization reaction it is necessary to avoid polymer degradation during the following processing steps. Therefore, the polymer solution, resulting from the polymerization, is first deactivated and stabilized and afterwards the polymer is separated from the solution and then characterized from a physico-mechanical point of view.

Stabilization of *cis*-1,4-polyisoprene is necessary because the polymer, having a high degree of unsaturation, is unstable toward a number of physical and chemical agents such as oxygen, ozone, light, high temperatures, UV radiations, etc. The above-mentioned agents can degrade the polymer by crosslinking and scission of the macromolecular chains [429]. The degradation reaction can be stopped by addition of certain chemical substances — called stabilizers — able to react with macroradicals to prevent crosslinking and chain scission reactions [430]. On the other hand, in order to prevent polymer degradation, the necessity for the deactivation of traces of active catalyst from the polymer solution becomes obvious, since they can not only cancel the effect of stabilizers, but even become initiators of the degradation process [429]. Therefore, in order to preserve the physico-chemical properties of *cis*-1,4-polyisoprene synthesized from both isoprene fractions, the need for polymer stabilization, deactivation and removal of traces of catalyst was obvious. The polymer was separated from solution, dried and its physico-mechanical properties were determined.

1. Stabilization and deactivation

A great number of stabilizers for synthetic rubbers is reported in literature, but only a few of them are used to stabilize polyisoprene. The choice of a stabilizer is made by taking into account all factors contributing to rubber degradation. It is known that a universal stabilizer which provides an identical protection against all factors producing rubber degradation is not available [431]. Therefore, investigations are directed to find stabilizers based on synergistic mixtures, trying to combine their various modes of action [432]. The stabilizers can be classified into two groups: stabilizers based on aromatic amines, highly coloured, but with a more efficient protective capacity than the second group of stabilizers based on phenols, which are non-staining [431].

Treatment with certain compounds, which results in the formation of acidic or basic products according to their nature, is recommended for the removal of traces of catalysts from the polymerization solution. Basic compounds such as ammonia, aliphatic or aromatic amines, etc., yield basic salts of aluminium or titanium or even hydrated aluminium oxide. Such products are insoluble and precipitate on the surface of the polymer particles resulting in a rubber with a high ash content [433]. Acidic type deactivation agents such as primary or secondary alcohols, acetone, etc., react in the presence of strong or weak acids, with traces of catalyst yielding soluble titanium and aluminium salts, e.g. chlorides. Such salts are easily removed from the polymer solution by repeated washing with water [434–436].

In the present work polymer stabilization and the deactivation of traces of catalyst was done concomitantly. The stabilization efficiency of several stabilizers was investigated. Among the staining stabilizers, the best results were given by *N*-phenyl-*N'*-cyclohexyl-*p*-phenylenediamine, while among the non-staining stabilizers, the most convenient was 4-methyl-2,6-di-*tert*butylphenol. Among the acidic deactivating agents, acetone was chosen for deactivation because it has a small molecule and can easily penetrate through the polymer macromolecules to provide a complete deactivation. The products are soluble and easily removed which results in a polymer with a low ash content. At the same time, acetone is also a good solvent for the stabilizing agents used. The polymer solutions obtained in the isoprene polymerization experiments were deactivated with acetone in a ratio of 22 moles/mole of catalyst (AlR_3). The stabilizer was added to the above amount of deactivating agent in a quantity of 1–2 % based on the polymer. After deactivation and stabilization, the catalyst residues were removed by washing with very dilute solutions of hydrochloric acid and then with distilled water. Complete residue removal was confirmed by determining the ash content of the polymer after each washing. The process was complete when the wash water was no longer acidic and the ash content of the polymer was below 0.2 g/100 g polymer.

The efficiency of the stabilizer was followed on unvulcanized rubber samples under normal temperature and pressure conditions for a period of 12 months by measuring, at given intervals of time, the intrinsic viscosity, the gel content and the Mooney viscosity. The physio-mechanical properties of the polymer were determined at the same intervals of time on vulcanized rubber samples. The above-mentioned stabilizers were found adequate, preserving the polymer characteristics during the entire period investigated.

2. *Separation of polymer*

The separation of polymer from solution was made after stabilization and deactivation in warm water, using methylcellulose as a dispersing agent.

Separation of polymer was carried out by gradual introduction, with stirring, of the polymer solution into distilled water, made alkaline with ammonia solution and containing methylcellulose as the dispersing agent. The water was heated to a temperature higher than the boiling point of the solvent. After evaporation, the solvent was condensed and separated from the polymer. Polymer particles rose to the surface of the water because of the difference in density. The removal of the suspended stabilizer from the surface of the polymer particles was achieved by repeated washing with distilled water. The resulting polymer was then dried, after squeezing out much of the water on a mill.

3. *Processing of* cis-*1,4-polyisoprene*

Polymers synthesized from both isoprene fractions, after deactivation, stabilization and separating from the polymerization solution, were characterized from a physico-mechanical point of view by measuring their tensile strength, elongation at break, modulus of 300% and yield point. A standard batch recipe and constant vulcanization conditions were used.

In order to gather comparative data on the physico-mechanical properties of the polymers tested, a sample of standard *cis*-1,4-polyisoprene type I, with similar physico-mechanical properties was also tested under the same conditions. The characteristics of the polymer are given in Table 31.

Table 31
Properties of *cis*-1,4-polyisoprene

Property	*cis*-1,4-Polyisoprene from		Standard sample type I
	over 99% isoprene fraction	*ca.* 25% isoprene fraction	
Content of *cis*-1,4 units, %	94	93	93
Intrinsic viscosity, dl/g	4.2	4.1	4.3
Gel, %	12	11	10
Ash, %	0.2	0.2	0.3

From the data given above we concluded that the properties of the synthesized polymers were maintained during the stablization, deactivation and separation steps. No degradation and crosslinking occurred, as shown by the low gel content and the intrinsic viscosity data. The low ash content of the polymers indicates that the deactivation and the removal of traces of catalyst were successful. The content of *cis*-1,4-structural units remained constant, as well.

The physico-mechanical tests were made on vulcanized rubber samples prepared with the following standard recipe:

Compound	Amount, w/w
1. Rubber	100
2. Stearic acid	1
3. Zinc oxide	5
4. Antioxidant 4010	1
5. Vulcacit DM (MBTS)	1
6. Vulcacit D (DPG)	3
7. Sulphur	2.5

Mixing temperature 50–60°C.

Vulcanization of the rubber mixes was carried out for 10 and 20 minutes, at both 125 and 135°C. The physico-mechanical properties after vulcanization are listed in Table 32.

Table 32

Physico-mechanical properties of polyisoprene obtained from over 99% and *ca.* 25% isoprene fraction and of *cis*-1,4-polyisoprene rubber (type I)

No.	Polymer	Vulcanization temperature (°C)	Vulcanization time (min)	Tensile strength (kgf/cm^2)	Elongation at break (%)	300% Modulus (kgf/cm^2)	Set at break (%)
1	Polyisoprene from over 99% isoprene	125	10	251	735	12	9
2			10	255	725	13	10
3			20	265	710	14	8
4			20	268	705	14	8
5		135	10	280	690	16	8
6			10	270	700	14	8
7			20	256	715	16	8
8			20	252	730	15	9
9	Polyisoprene from *ca.* 25% isoprene	125	10	253	728	13	9
10			10	252	730	12	10
11			20	260	720	14	8
12			20	264	715	13	8
13		135	10	270	700	15	9
14			10	278	690	15	10
15			20	253	718	14	10
16			20	250	725	16	11
17	*cis*-1,4-Polyisoprene type I	135	10	282	695	13	9
18			10	280	690	14	9
19			20	263	705	14	10
20			20	258	715	15	10

From data recorded on the physico-mechanical properties of *cis*-1,4-polyisoprene, we concluded that the polymers have high values of physico-mechanical properties. The values are very close for the polyisoprenes prepared from both isoprene fractions and are comparable with the reference sample.

As is well-known, the physico-mechanical properties of *cis*-1,4-polyisoprene are dependent on the physico-chemical properties, on the recipe used in processing and on the curing conditions. The high values of the physico-mechanical properties confirm the high physico-chemical properties of the synthesized polymers, the use of an adequate recipe for vulcanization and adequate curing conditions.

In order to confirm the properties of the *cis*-1,4-polyisoprene, sufficient polymer to manufacture a number of tyres was synthesized on a pilot plant. Both the tread and the tyre carcass were manufactured from a mixture of *cis*-1,4-polyisoprene and Carom 1500 rubber according to current manufacturing specifications for car tyres. During manufacture, it was established that the carcass and the treads showed good adherence to each other. After prolonged production trials, we concluded that the mixes for tread and carcass corresponded to the standards and showed no risk of scorch during milling and extrusion. The manufactured tyres were mounted on a car and tested on various road surfaces.

Conclusions

IN THE present work, the results of a research undertaken with a view to finding experimentally a correlation between the types of isoprene fraction, the catalyst system, the reaction conditions and the properties of the resulting polyisoprene are reported as well as some aspects of the kinetics of the polymerization reaction.

In the experimental part the following questions were studied:

I. Purification of the reagents used in stereospecific polymerization in order to establish the conditions of reagent purity necessary to achieve reproducible polymerization reactions.

II. Investigation of the stereospecific polymerization of isoprene fractions of over 99% purity and *ca.* 25% purity with heterogeneous catalyst systems, namely $Al(C_2H_5)_3 + TiCl_4$ and $Al(i\text{-}C_4H_9)_3 + TiCl_4$ in order to obtain a polyisoprene with a high content (94%) of *cis*-1,4-structural units with physico-chemical and physico-mechanical properties similar to those of natural rubber, using different solvents, and catalyst systems of differing activity and stereospecificity.

III. Some aspects of the kinetics of the stereospecific polymerization of isoprene including the influence of temperature and polymerization time on the rate of the polymerization reaction and the molecular weight of the polymer.

IV. IR absorption spectra, degree of unsaturation, crystallinity and molecular weight distribution of the *cis*-1,4-polyisoprene were determined to compare the properties of the polymers synthesized from each of the isoprene fractions with a standard *cis*-1,4-polyisoprene used as a reference.

V. Optimum conditions were established for the stabilization, deactivation and the removal of solvent from polyisoprene solutions in order to preserve the properties of the resulting polymer as well as the optimization of a mix recipe and a curing procedure to give vulcanizates with improved physico-mechanical properties.

VI. Synthesis of a larger quantity of *cis*-1,4-polyisoprene for the manufacture of car tyres for testing on various road surfaces.

The following conclusions can be drawn from the experimental results:

1. It was found that impurities such as the cyclopentadiene, carbonyl compounds and acetylenic hydrocarbons present in isoprene, and the water and oxygen present in the isoprene, the solvent and the nitrogen must not exceed a concentration of 10 ppm of each component in order to achieve reproducible polymerizations. Since saturated hydrocarbons are used as solvents in the stereospecific polymerization of isoprene, the isoprene fraction of about 25 % purity could also be used for polymerization. Investigations on the polymerization of isoprene with a purity of about 25 %, the other components being 63 % of isopentane and 11 % of olefins, have not been made before. It is thus possible to avoid certain costly concentration and purification steps in the synthesis of isoprene.

2. Procedures for the removal of the above-mentioned impurities from isoprene, solvents and nitrogen, down to the permitted limits were worked out:

— cyclopentadiene, by treating the isoprene fractions with maleic anhydride,
— oxygen from isoprene and solvent, by distillation, and from nitrogen, by using a copper catalyst,
— water, carbonyl compounds and acetylenic hydrocarbons, by passage through molecular sieves.

3. Isoprene must be submitted to deaeration and drying before use; the storage of isoprene for 24 to 92 hours after purification produces a gradual decrease of the reaction rate, proportional to the storage time.

4. The use of a hexane fraction with a boiling range between 65 and 72°C permits easy removal of the solvent from the polymer.

5. From the investigation of the polymerization of both isoprene fractions with the $Al(C_2H_5)_3 + TiCl_4$ and $Al(i\text{-}C_4H_9)_3 + TiCl_4$ catalyst systems the following conclusions were made:

(a) Polyisoprene with good physico-chemical and mechanical properties similar to those of natural rubber, are prepared from the isoprene fractions under virtually identical conditions, namely a monomer/solvent ratio of 1/4 (v/v), and $AlR_3/TiCl_4$ molar ratio of 1, an AlR_3 concentration of 1 g/100 g isoprene, a temperature of 30–32°C and a polymerization time of 120 minutes.

The polymerization of an isoprene of *ca.* 25 % purity with heterogeneous catalyst systems has not been reported before in the literature and was therefore patented. The method is also of theoretical interest, since in the given reaction conditions it performs a selective polymerization of isoprene mixed with saturated and unsaturated hydrocarbons (mono-olefins).

(b) The maximum activity and stereospecificity of the two catalyst systems for the synthesis of *cis*-1,4-polyisoprene are governed by the Al/Ti molar ratio and by the concentration of the catalyst.

— The stereospecificity reaches a maximum at an Al/Ti molar ratio of 1 or above, the resulting polymer containing about 94% *cis*-1,4-addition product. As the Al/Ti molar ratio decreases to values less than unity, e.g. 0.5, the polymers contain only 30% of *cis*-1,4 product, and 50% *trans*-1,4 product, the remainder being products of 1,2- and 3,4-addition. The stereoregularity is independent of the nature of the alkyl group, the catalyst concentration, the temperature and the reaction time within the range investigated. The nature of the solvents used in this work (*n*-heptane, the hexane fraction and isopentane) does not affect the microstructure of the polymer.

— The catalyst activity is a maximum at an Al/Ti molar ratio of 1, but is dependent on its concentration and polymerization temperature. At an Al/Ti molar ratio of 0.5, a granular polymer with low molecular weight (intrinsic viscosity = 0.2 dl/g) is obtained; at an Al/Ti molar ratio of 1, the molecular weight reaches its maximum value (intrinsic viscosity = 4.5–5 dl/g), and as the Al/Ti molar ratio increases, the molecular weight decreases, the formation of polymers with a low viscosity and high content of oily products being favoured.

— The dependence of the polymerization reaction rate on the Al/Ti molar ratio is similar to the relationship shown by the molecular weight. Maximum rate occurred at an Al/Ti molar ratio of unity.

The different reactivities of the catalyst complex prepared with various Al/Ti molar ratios may be explained by the degree of reduction of the transition metal.

At an equimolar ratio between trialkylaluminium and titanium tetrachloride, the transition metal is completely reduced from tetravalent titanium to trivalent titanium; the resulting complex has the maximum activity and stereospecificity, favouring the formation of *cis*-1,4-polymer. In the catalyst complex formed at an Al/Ti molar ratio of 0.5, the titanium is only partially reduced from tetravalent titanium to trivalent titanium because there is insufficient aluminium alkyl present in the system and therefore it still contains titanium in its highest valency state; such a catalyst complex results in the formation of crosslinked polymers with low stereospecificity and molecular weight. The rate of the polymerization reaction is also slow.

At Al/Ti molar ratios higher than unity, the activity of the catalyst complex decreases again owing to the excess of aluminium alkyl.

(c) The catalyst concentration also influences the rate of the polymerization reaction and the molecular weight, the other reaction variables being kept constant. The highest molecular weight was obtained at a concentration of 1 g AlR_3/100 g isoprene; the reaction rate was proportional to the catalyst concentration. Although at a concentration of 1.5 g AlR_3/100 g isoprene, the reaction rate is somewhat faster than at a concentration of 1 g AlR_3/100 g isoprene, the latter produces the most convenient values of molecular weight, which is an important factor in the usage of polyisoprene.

(d) In this research, no difference was observed as to the influence of the nature of the alkyl group of the aluminium alkyls on the $AlR_3/TiCl_4$ molar ratio, its concentration or on the characteristics of the polymer. Differences in the rates of the polymerization were, however, observed. Under similar reaction conditions, isoprene polymerizes faster in the presence of the $Al(i\text{-}C_4H_9)_3 + TiCl_4$ catalyst stem than in the presence of the $Al(C_2H_5)_3 + TiCl_4$ system. From this point of view, the catalyst based on $Al(i\text{-}C_4H_9)_3$ is more advantageous and the differences produced in molecular weight, somewhat lower in this case, do not affect the values of the physico-mechanical properties.

The two catalyst systems have a similar activity in the polymerization of both the isoprene fractions.

6. The kinetics of the polymerization reaction are influenced by the following variables: the Al/Ti molar ratio, the concentration of the catalyst, the nature of the solvent, the temperature and polymerization time.

(a) The polymerization reaction of isoprene fractions of over 99% and *ca.* 25% purities takes place in the temperature ranges 10 to 55°C and 10 to 35°C, respectively. The rate increases with increasing temperature; under the same conditions, the molecular weight decreases. The most effective range of polymerization temperature for both isoprene concentrations was established to be 30–35°C.

(b) The reaction time necessary to reach a conversion higher than 80% is 120 minutes.

(c) The molecular weight increases in the first stage of the reaction, and reaches the highest value after 30 minutes and 60 minutes, respectively, thereafter remaining constant during the interval investigated, i.e. 300 minutes.

(d) Under constant working conditions, the microstructure of the polyisoprene is not changed during the polymerization reaction.

7. IR absorption spectra within the 700 to 4000 cm^{-1} spectral range, the degree of unsaturation and the degree of crystallinity of polyisoprene prepared from both isoprene fractions are not different from those of a reference *cis*-1,4-polyisoprene. This is a further evidence that the isoprene fraction of *ca.* 25% purity can be converted into *cis*-1,4-polyisoprene with improved elastomeric characteristics. Polyisoprene obtained from both isoprene fractions has the wide molecular weight distribution which is specific for the catalyst systems used.

8. Methods have been devised for the stabilization, deactivation and removal of solvent from the polymer solutions, in order to preserve the properties of the synthetic polymers.

(a) The stabilizers giving the best results were *N*-phenyl-*N′*-cyclohexyl-*p*-phenylenediamine and 4-methyl-2,6-di-*tert*-butylphenol.

(b) Acetone, an acidic type deactivation agent, is successful in removing catalyst residues from the polymer.

(c) The separation of solvent from the polymer is effected by using water at temperatures depending on the boiling point of the solvent, with methylcellulose as a dispersing agent.

9. A standard recipe and vulcanization conditions were found applicable to the polyisoprene prepared from both isoprene fractions, as well as to a reference sample of *cis*-1,4-polyisoprene; similar physico-mechanical properties were obtained, comparable with those of natural rubber.

10. A larger amount of *cis*-1,4-polyisoprene was synthesized in a pilot plant. Tyres were manufactured, mounted on a car and tested by driving on various road surfaces at different times of the year. These tyres lasted for about 45,000 km, compared with the 30,000 to 35,000 km which is customary.

Polyisoprenes synthesized from both the isoprene fractions used have properties similar to those of a reference *cis*-1,4-polyisoprene sample and comparable with those of natural rubber.

PART II

Introduction

THE SYNTHESIS of *cis*-1,4-polyisoprene rubber is a most interesting and important subject, both theoretically and practically, which made us continue the research in this particular field.

Polyisoprene rubber, as a substitute for natural rubber, is needed in large and continuously increasing quantities within the framework of general development of the industry, as it is used in so many branches. For these reasons a technological research was initiated and the theoretical aspects of the subject have been studied at the same time.

The ultimate purpose of the research was to develop a manufacturing process serving as a basis for the design and building of a large industrial plant. Particular emphasis was also laid on the source of raw materials in order to achieve economical manufacturing of the isoprene monomer. An important source is the C_5 fraction from cracked gasoline. The development of the plastics industry has led to the building of a great number of pyrolysis plants for supplying the basis monomers, e.g. ethylene, propylene, butadiene, etc. A cracked gasoline is also produced with a great number of components, among which is isoprene, present in a low concentration of only 3–5%, depending on the raw material used in pyrolysis as well as on the operating conditions. In the C_5 cut isolated from cracked gasoline, isoprene is present in concentrations of 15–20%, together with other components such as cyclopentadiene, piperylene, etc., which can also be used to manufacture other types of elastomers. Isoprene with a concentration of over 99%, with a very low content of impurities, is obtained from the C_5 cut by physical processes of separation and purification. This product can be used for stereospecific polymerization.

Isopentane has been chosen as a polymerization solvent for economic reasons. At equal concentrations of polyisoprene, solutions with lower viscosity are obtained in isopentane, as compared with other solvents, while the energy consumption to remove the solvent from polymer solutions is less, owing to the lower boiling temperature of isopentane.

Special efforts were made to improve the catalyst system as it is a key factor in the polymerization reaction, determining the polymer characteristics and the level of unit consumptions of the catalyst. An attempt has been made to find a correlation between some of the properties of the catalyst complexes and their activity in polymerization, by employing modern techniques for their characterization.

Some practical and theoretical aspects of the solution polymerization of isoprene were studied with a special request for the purity necessary in the polymerization medium. Assumptions on gel formation and structure and on its influence on the properties of the synthetic polyisoprene have been formulated.

Research on the deactivation of the catalyst complex, and the stabilization and separation of the polymer from solution, was undertaken in order to obtain *cis*-1,4-polyisoprene with improved properties. Highly accurate modern methods were used to determine some properties of the *cis*-1,4-polyisoprene.

Chapter I

Pre-formed Ziegler–Natta-type catalyst systems

AS PREVIOUSLY indicated, Ziegler–Natta-type catalysts are complex systems formed by the reaction between an organometallic compound of groups I–III (Li, Na, Mg, Ca, Be, Zn, Al) and a compound (usually a halide or an oxyhalide) of a transition metal of groups IV–VIII (Ti, V, Cr, Ni, W, etc.).

The catalyst system of practical interest in the stereospecific polymerization of isoprene to give *cis*-1,4-polyisoprene contains $TiCl_4$ and a trialkylaluminium [1–3]; the most used one is triisobutylaluminium (TIBA) [4]. By reaction between the above two components at convenient molar ratios, a reddish-brown precipitate is formed which contains mainly of β-$TiCl_3$. This is the active component in a heterogeneous Ziegler–Natta-type catalyst complex catalysing the stereospecific polymerization of isoprene.

Formation of $TiCl_3$ most probably takes place by the alkylation of titanium halide with aluminium alkyl, followed by a homolytic decomposition of the Ti^{4+} organic derivative, according to the following reaction sequence [5]:

$$TiCl_4 + AlR_3 \rightarrow RTiCl_3 + AlR_2Cl \quad (1)$$

$$TiCl_4 + AlR_2Cl \rightarrow RTiCl_3 + AlRCl_2 \quad (2)$$

$$TiCl_4 + AlRCl_2 \rightarrow RTiCl_3 + AlCl_3 \quad (3)$$

$$RTiCl_3 \rightarrow TiCl_3 + R\cdot \quad (4)$$

The β-form of titanium trichloride consists of a hexagonal unit cell containing two $TiCl_3$ monomeric units.

It should be noticed that in the reduction of $TiCl_4$ with trialkylaluminium according to the above scheme, the resulting β-$TiCl_3$ contains aluminium as aluminium chloride, present as a solid solution imbedded into the crystal lattice via an isomorphous substitution of titanium atoms with the aluminium atoms during the formation of the crystal [6].

The lattice defects of β-$TiCl_3$ are actually the active sites in polymerization [7], which emphasizes that the catalytic activity of such stereospecific catalysts is due to both the chemical nature and the physical structure of the system.

Hence, the conditions of formation of a Ziegler–Natta-type catalyst complex, i.e. either *in situ* (in the presence of monomer) or *pre-formed* (in the

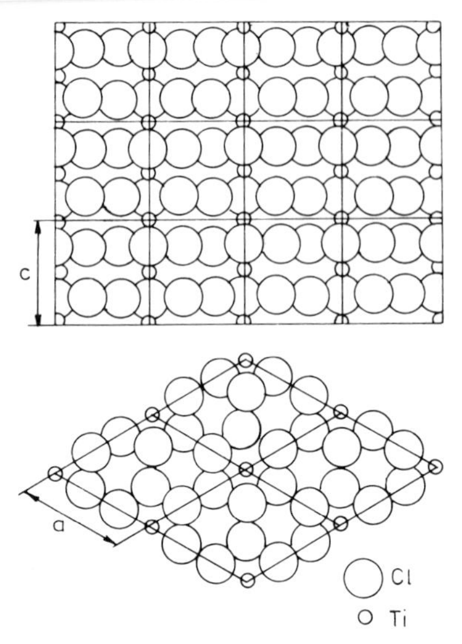

FIGURE 1. Crystalline structure of β-$TiCl_3$ [6].

absence of monomer), as well as other working variables, such as temperature, stirring, etc., have a major influence on the complex characteristics and hence on the course of the polymerization reaction and on the properties of the polymer.

1. Two-component AlR_3 + $TiCl_4$ catalyst systems

Ziegler–Natta catalyst systems can be divided into two types, i.e. prepared *in situ* and pre-formed, depending on the presence or absence of the monomer during the preparation of the catalyst complex. Pre-formed Ziegler–Natta-type catalyst complexes were used in order to increase their catalytic activity, to reduce the unit consumption of the catalyst and to improve the properties of the *cis*-1,4-polyisoprene. The versatility of the polymerization conditions is enhanced by pre-forming the catalyst complex, since it can be prepared at different temperatures and concentrations and in solvents different from those of the actual polymerization medium.

A comparison between the two catalyst types, i.e. *in situ* and pre-formed, is made in Table 1. The data listed in Table 1 confirm the higher reactivity of pre-formed catalyst complexes as compared to the ones prepared *in situ* [8].

In the polymerization of isoprene with catalyst complexes prepared *in situ* a short induction period is observed, which disappears when pre-formed complexes are used (Figure 2).

Table 1
Effect of TIBA-$TiCl_4$ complex preparation on isoprene polymerisation [8]

Catalyst concentration, 10^{-4} moles $TiCl_4$	Yield of solid polymer, %		Properties of polyisoprene $[\eta]$, dl/g		Gel, %	
mole isoprene	I*	II**	I*	II**	I*	II**
0.18	46	27	5.7	5.2	13	27
0.27	69	43	5.5	5.2	18	12
0.36	76	57	4.8	4.7	12	7
0.89	87	74	3.9	4.0	9	10

Solvent of polymerisation and preparation of the unmaturated complex, heptane: Al/Ti molar ratio = 1.

* Fresh catalytic complex preformed at room temperature.

** Catalytic complex prepared *in situ*.

Pre-forming of the catalyst complex in appropriate conditions (solvent, temperature, etc.) results in a better reproducibility of the polymerization reaction and yields a *cis*-1,4-polyisoprene with a higher molecular weight, a lower gel content and less extractable material (i.e. products with a relatively low molecular weight), than is obtained by polymerization with catalyst complexes prepared *in situ*.

Based on the above-mentioned advantages, modern manufacturing processes of *cis*-1,4-polyisoprene use only either two-component or modified pre-formed Ziegler–Natta-type catalyst complexes.

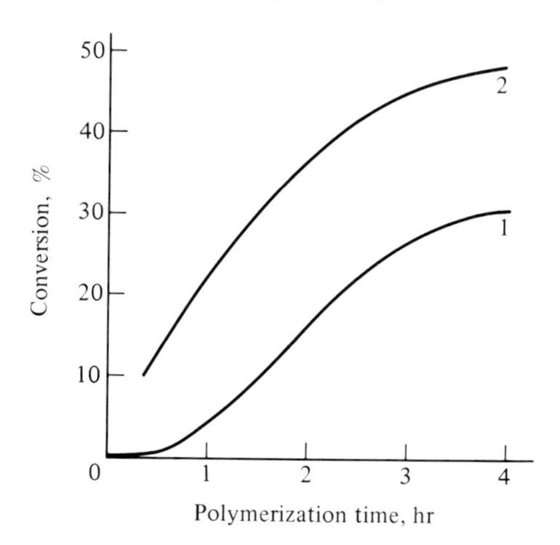

FIGURE 2. Plot of isoprene conversion against time in polymerization with TIBA + $TiCl_4$ catalyst complex prepared *in situ* (curve 1) and pre-formed (curve 2). Catalyst concentration = 0.18×10^4 mole $TiCl_4$/mole of isoprene [8].

Pre-formed heterogeneous Ziegler–Natta-type catalyst complexes are prepared by the reduction reaction of $TiCl_4$ with AlR_3 in the presence of a hydrocarbon as a reaction medium. The properties of the resulting catalyst complex are strongly dependent on the preparation conditions. The most important factors which influence the catalyst complex activity are: the pre-forming solvent, the order in which the catalyst components are added, the pre-forming temperature, the ratio of the catalyst components and the period of catalyst maturation.

All the above-mentioned factors which influence the catalyst activity, will be discussed below.

1.1. The pre-forming solvent

As a reaction medium for Ziegler–Natta-type catalyst complexes, non-polar and chemically inert compounds are used such as straight or branched-chain paraffins, cycloalkanes and aromatic hydrocarbons. The catalyst components are dissolved in the above solvents and finally the catalyst complex is obtained as a slurry [8–14].

For technological reasons both the pre-forming and the polymerization steps were first carried out in isopentane in order to simplify the solvent-recovery operation.

In comparative studies on the polymerization of isoprene with a catalyst complex based on pre-formed TIBA + $TiCl_4$ in isopentane and toluene (Table 2), higher molecular weights and lower gel content of the polymer, as shown by higher Mooney plasticities, were obtained with the latter solvent.

The data shown in Table 2 are in agreement with literature data which report that catalyst complexes pre-formed in aromatic hydrocarbons, especially toluene and benzene, result in the formation of a polyisoprene with a

Table 2

Effect of the nature of the preforming medium on isoprene polymerisation with TIBA + $TiCl_4$ catalyst system

[catalyst concentration 1.5 g $TiCl_4$/100 g isoprene]

Preforming medium	Preforming temperature, °C	Conversion (after 4 h) %	Polyisoprene characteristics		
			$[\eta]^{(a)}$, dl/g	Gel, %	ML(1 + 4) at 100°C
Isopentane	20	86	3.91	26	56
	30	74	3.87	25	56
Toluene	20	95	4.54	17	95
	30	90	4.10	15	83

(a) In toluene at 30°C.

lower gel content and higher molecular weight than those pre-formed in saturated hydrocarbons [11]. This effect can be explained by the complexes formed between $TiCl_4$ and aromatic hydrocarbons [11]. Such donor–acceptor type complexes are coloured, the aromatic hydrocarbon supplying the vacant d^2sp^3-orbitals of titanium with π-electrons from their extended orbital [15]. Different coloured complexes are obtained depending on the aromatic solvent. The $TiCl_4$–benzene complex is yellow, the $TiCl_4$–xylene and $TiCl_4$–toluene complexes are orange, while the $TiCl_4$–diphenyl complex is scarlet.

Catalyst complexes pre-formed in toluene were stable over a longer period of time than when isopentane was used as a pre-forming solvent.

1.2. Order of adding the catalyst components

The catalyst complex activity, as expressed by the properties of the polyisoprene obtained in polymerization with it, strongly depends on the order of adding the solutions of the two catalyst components. There are three possibilities:

(a) addition of aluminium alkyl solution to $TiCl_4$ solution;
(b) addition of $TiCl_4$ solution to aluminium alkyl solution;
(c) simultaneous mixing of both solutions of the catalyst components.

Ziegler–Natta-type catalyst complexes, with different activities in isoprene polymerization, are obtained according to the mixing method chosen. The differences in activity are produced by the original excess of one of the catalyst components. A catalyst prepared by addition of a $TiCl_4$ solution to aluminium alkyl solution is less active. This was attributed to the excess of aluminium alkyl which may behave in two ways:

— it is adsorbed on the β-$TiCl_3$ crystal surface and blocks the active polymerization sites;
— it reacts with $TiCl_3$, further reducing it to $TiCl_2$ according to the reaction

$$TiCl_3 + AlR_3 \rightarrow TiCl_2 + AlR_2Cl + R\cdot \qquad (5)$$

and the $TiCl_2$ is no longer active in isoprene polymerization.

Complexes resulting from the addition of aluminium alkyl solution to $TiCl_4$ solution yield a polyisoprene with a somewhat higher gel content. A possible explanation could be that some $TiCl_4$ remains in its unreduced form and this is able to initiate cationic polymerization reactions resulting in the formation of crosslinked polymers.

The intrinsic viscosity and microstructure of the sol fraction of polyisoprene are not affected by the order in which the catalyst components are added, provided the optimal Al/Ti molar ratio corresponding to both the order of addition and the pre-forming temperature was used.

Best results were obtained with catalyst complexes prepared by simultaneously mixing the catalyst components in the pre-forming reactor. However, for laboratory runs, the first procedure was chosen, i.e. addition of aluminium alkyl solution to $TiCl_4$ solution.

The optimal Al/Ti molar ratio too, is influenced by the order of adding the components, but it also depends on the pre-forming temperature as will be discussed in Section 1.3.

1.3. Temperature of pre-forming of complex

The temperature at which the catalyst complex is prepared plays a major role in its activity. The lower the pre-forming temperature, the more active the catalyst complex. The data listed in Table 3 emphasize the dependence of the activity of the catalyst complex on the pre-forming temperature in the case of TIBA + $TiCl_4$.

As shown in Table 3, a decrease in the pre-forming temperature from +30° to −40°C causes the rate of the polymerization reaction to reach about double its original value at catalyst concentrations 2.5 times as low. The physico-chemical properties of *cis*-1,4-polyisoprene (intrinsic viscosity, gel content, tacticity) are not affected by the pre-forming temperature.

The marked increase in catalyst activity as a result of decreasing the pre-forming temperature is explained by the formation of a catalyst complex with a very small particle size and hence a high surface area, owing to the slower reduction of the $TiCl_4$ by the aluminium alkyls. This results in the formation of a greater number of reactive polymerization sites, since it is known that the number of active polymerization sites is proportional to the surface area of the catalyst complex. Moreover, catalyst complexes pre-formed at lower temperatures display an improved stability in time over the ones pre-formed at temperatures higher than 0°C.

1.4. Molar ratio of the components of the catalyst

As mentioned above (Section 1.2) the optimum $TiCl_4$/TIBA molar ratio leading to a catalyst complex with a maximum activity and stereospecificity in the polymerization of isoprene depends on the order in which the catalyst components are mixed and on the pre-forming temperature.

When the catalyst components are added simultaneously to the pre-forming system, the optimum value was found to be the equimolar ratio (Figure 3). When the $TiCl_4$ solution is added to TIBA solution, the optimal Al/Ti molar

Table 3

Effect of pre-forming temperature on the reactivity of the TIBA + $TiCl_4$ catalyst complex in the polymerization of isoprene

No.	Pre-forming temperature (°C)	Catalyst concentration g $TiCl_4$/100 g isoprene	Polymerization time (hr)	Conversion (%)	Polyisoprene characteristics		
					$[\eta]$[a] (dl/g)	Gel (%)	Tacticity *cis*-1,4 (%)
1	−40	0.6	2	98	4.67	17	96.3
2	−30	0.6	2	90	4.62	18	94.6
3	−20	0.6	2	87	4.67	16	95.8
4	−10	0.75	2	90	4.72	16	95.9
5	0	0.75	2	75	4.58	17	95.9
6	10	1.5	2	97	4.61	15	96.1
7	20	1.5	4	95	4.54	17	95.7
8	30	1.5	4	90	4.10	16	96.3

[a] In toluene at 30°C.

Catalyst complex pre-formed in toluene and maturated for 1 hour at room temperature; Al/Ti molar ratio = 1.

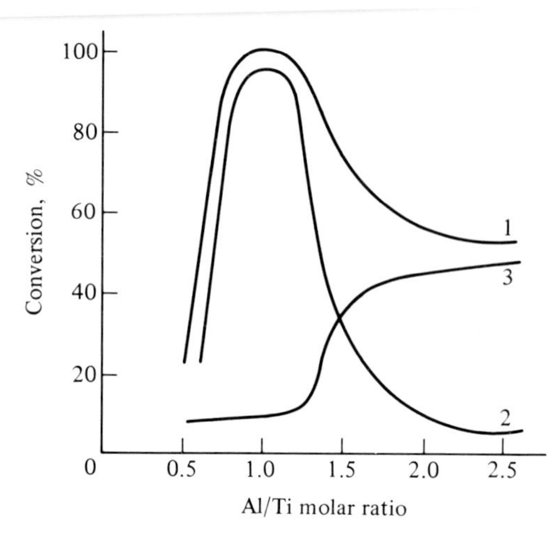

FIGURE 3. Optimal Al/Ti molar ratio in pre-forming the catalyst complex by simultaneous addition of catalyst components. 1. Total polyisoprene; 2. solid polymer; 3. extractables [8].

ratio shifts towards values higher than unity and increases with the pre-forming temperature (Figure 4, curve 1) without exceeding 1.5 [8, 9].

When TIBA solution is added to a $TiCl_4$ solution at low temperatures, the optimum Al/Ti molar ratio also takes values about unity (Figure 4, curve 2), or less at higher temperatures.

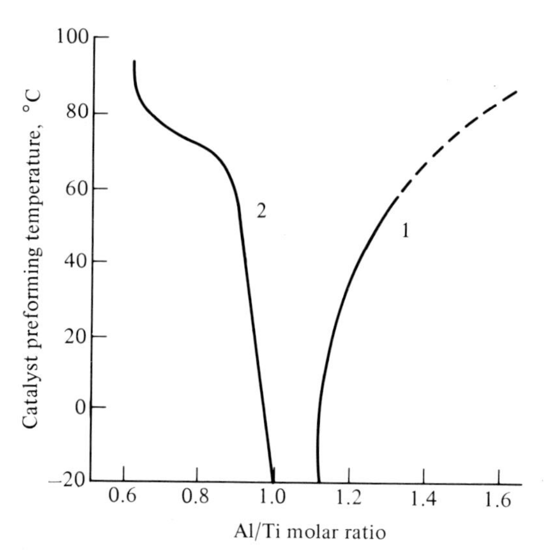

FIGURE 4. Dependence of pre-forming temperature on the optimum Al/Ti molar ratio. 1. Addition of $TiCl_4$ solution to TIBA solution; 2. Addition of TIBA solution to $TiCl_4$ solution [8].

Moreover, ESR studies of a catalyst complex containing TIBA and $TiCl_4$, pre-formed in toluene at $-20°C$, by adding the TIBA solution to the $TiCl_4$ solution, indicate the presence of a maximum concentration of unpaired electrons, corresponding to the active species, at an equimolar ratio of the components as shown in Section 1.6, Figure 8.

1.5. Maturation of the catalyst complex

Although the reduction reaction of $TiCl_4$ with TIBA is very fast, a certain period of time should be allowed for its completion that the maximum number of active polymerization sites is formed. This is called the *maturation time.*

In studies on isoprene polymerization with unmaturated (freshly prepared) and maturated Ziegler–Natta-type catalysts, it was found that the maturation of the catalyst complex influences, to a large extent, both the polymerization reaction and the polymer characteristics.

Low molecular weight fractions (extractables) were found in polyisoprene synthesized with a fresh (unmaturated) catalyst, whereas the polymer synthesized with maturated catalysts, contained virtually no extractable fractions (Table 4).

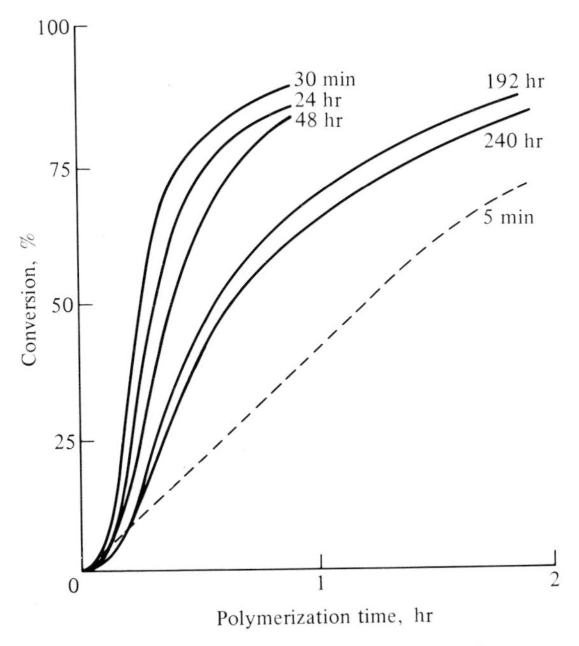

FIGURE 5. Effect of maturation time on activity of TIBA + $TiCl_4$ catalyst system as expressed by the degree of conversion in isoprene polymerization with a Al/Ti molar ratio of 1 [12].

Table 4

Effect of the maturation time on the reactivity in isoprene polymerization of the TIBA + $TiCl_4$ catalyst complex pre-formed in toluene

No.	Pre-forming temperature (°C)	Maturation time (min)	Polymerization time (hr)	Catalyst concentration g $TiCl_4$/100 g isoprene	Conversion (%)	Polymer characteristics		
						$[\eta]$[a] (dl/g)	Extractables (%)	Tacticity *cis*-1,4 (%)
1	20	0	4	1.5	62	3.89	5.8	95.4
2	20	60	4	1.5	95	4.54	2.8	95.7
3	−20	0	2	0.6	56	3.96	4.7	95.4
4	−20	60	2	0.6	89	4.67	0.9	95.8

[a] In toluene at 30°C.
Al/Ti molar ratio = 1.

The polymerization rate of isoprene initially increases with maturation time and then decreases (Figure 5). In longer maturation times, the so-called "ageing" of the catalyst complex occurs [12].

ESR studies of the TIBA + $TiCl_4$ catalyst system, pre-formed in toluene at $-20°C$, showed that a variation with time of the concentration of unpaired electron (regarded as being the species active in isoprene polymerization) occurred. At 25°C a maximum concentration is reached about 8 hours after the mixing of the reagents, remaining constant for a further 12–14 hours and then decreasing (Figure 6). Hence *ageing* of Ziegler–Natta-type catalyst complex containing TIBA + $TiCl_4$ starts 20–24 hours after its preparation. This process of catalyst ageing can be slowed down by stirring the catalyst complex at low temperatures (about $-20°C$).

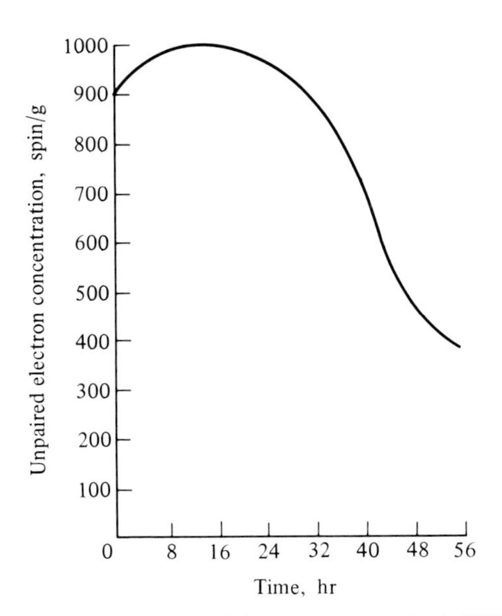

FIGURE 6. Effect of storage time on unpaired electron concentration in TIBA + $TiCl_4$ catalyst system with a Al/Ti molar ratio of 1.

The above results are in agreement with data on the dependence of the catalyst activity on the storage time.

The influence of the maturation temperature is also made evident by the fact that the maximum catalystic activity of the complex is more rapidly reached as temperature increases. This property of the catalyst complex cannot be used in practice, as it is associated with catalyst *ageing* which also occurs more rapidly at higher storage temperatures.

1.6. Electron spin resonance studies on a Ziegler–Natta-type catalyst complex

Ziegler–Natta-type catalyst systems are very complex and numerous approaches were made to find analytical methods able to provide *a priori* data on their catalytic activity. Such attempts also include electron spin resonance measurements (ESR).

An inspection of ESR spectra provides information on certain important physical properties of the system studied. The spectroscopic splitting g-factor indicates the degree of delocalization of unpaired electrons and local field symmetry. The area of the absorption curve is proportional to the number of paramagnetic centres in the analysed sample. The shape of the curve provides information on the distribution of paramagnetic centres and lattice defects. The width of the resonance line is related to interactions between the paramagnetic centres and the spin-relaxation time. The number of resolved lines in a spectrum and the ratio of their intensities provide data on the fine and hyper-fine interactions. In this way, the energy-level structure of the paramagnetic centre and the type of atomic nuclei adjoining the analysed centre can be determined [16].

(*a*) *General aspects*

Based on ESR studies of catalyst systems prepared by the reaction of $(C_5H_5)_2TiX_2$ (X = Cl, Br, I) with $Al(C_2H_5)_3$, $Al(i\text{-}C_3H_7)_3$, $Al(i\text{-}C_4H_9)_3$ or $Al(C_2H_5)_2Cl$, respectively, Shilov *et al.* [17, 18] formulated the concept of Ti–Al bimetallic complexes carrying unpaired electrons. The resonance active sites are identified by the $3d^1$-electrons of the Ti^{3+} ions. The resonance spectra of products isolated in the reaction of bis(cyclopentadienyl)titanium dichloride with $Al(i\text{-}C_4H_9)_3$ were similarly explained [19]. Maki and Randall [20] studied the ESR spectra of products isolated in the reaction of bis(cyclopentadienyl)titanium dichloride with various aluminium alkyls, e.g. $Al(CH_3)_3$, $Al(C_2H_5)_2Cl$, $Al(C_2H_5)Cl_2$, $Al(i\text{-}C_4H_9)_3$, $Al(C_2H_5)_3$. The authors concluded that the distribution of the unpaired electron adjacent to an aluminium atom changes with the nature of the substituents on the Ti^{3+} ion, the main role being ascribed to the inductive effect of chlorine atoms.

Adema *et al.* [21, 22] studied by ESR a mixture of $TiCl_4$ and either $Al(C_2H_5)_2Cl$ or $Al(C_2H_5)Cl_2$ in heptane. A correlation between the rate of polymerization and the number of unpaired electrons was found. Measurements were later made on other systems too, such as

$(C_5H_5)_2TiCl_2 + Al(CH_3)_3$, $(C_5H_5)_2TiCl_2 + Al(CH_3)_2Cl$,
$(C_5H_5)_2TiCH_3Cl + AlCH_3Cl_2$, $(C_5H_5)_2TiCl_2 + AlCH_3Cl_2$,
$(C_5H_5)_2TiCH_3Cl + AlCl_3$, at room temperature [23].

They suggest the following "structure", which carries an unpaired electron:

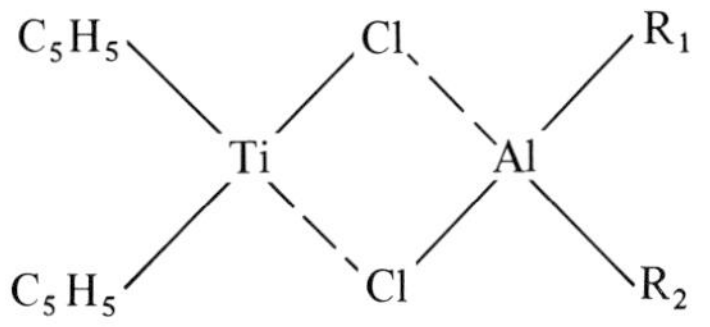

```
C5H5\        Cl\        /R1
     \      /   \      /
      Ti          Al
     /      \   /      \
C5H5/        \Cl        \R2
```

where R_1 and R_2 can be either methyl groups or chlorine atoms.

ESR spectra seem not to be influenced by C_5H_5 groups, which means that the unpaired electron density is relatively low in the proximity of those groups. Spin density must be higher in the proximity of Ti, Cl and Al atoms. In such a case g-values must depend on the substituent nature. Experimentally it was found that by changing the environment of the titanium ion, the g-values decrease with an increase of the chlorine content.

ESR studies of the reaction between $(C_5H_5)_2TiCl_2$ and $Al(CH_3)_2Cl$ in benzene solution were reported by Sylov *et al.* [24]. The reaction involves the formation of the $(C_5H_5)_2TiCH_3Cl{\cdot}Al(CH_3)Cl_2$ complex (denoted by A). Further reduction of this complex occurs only in the presence of olefins, according to the equation:

$$2A + RCH{=}CH_2 \rightarrow RC(CH_3){=}CH_2 + CH_4 + 2(C_5H_5)_2TiCl{\cdot}Al(CH_3)Cl_2 \qquad (6)$$

The characteristic g-value of the complex is similar to the one found when olefins are present; in the latter case the spectrum shows a well-resolved hyperfine structure of six equally intense lines, which emphasize an interaction of the unpaired electron with the ^{27}Al nucleus (with a nuclear spin $I = 5/2$).

Tkač [25] analysed the ESR spectrum of Ziegler–Natta-type catalyst systems under various conditions of reduction of $TiCl_4$ with $Al(i\text{-}C_4H_9)_3$ in hexane solution. Four types of paramagnetic centres were found in the separated phases of the heterogeneous catalyst systems. The absolute number of single types of active centres depends on the Al/Ti molar ratio, the temperature of pre-forming, and the order of mixing the components, as well as on the traces of oxygen and moisture present in the system. The stepwise replacement of chloride ions in the $TiCl_3$, combined with the Al in the bimetallic complex, by one or two alkyl groups, produces an increase in the g-value.

For a $Ti(OC_4H_9)_4 + Al(C_2H_5)_2Cl$ homogeneous catalyst at an Al/Ti molar ratio of 2, Angelescu *et al.* [26] obtained a spectrum with a resolved structure of eleven components assigned to the interaction of the unpaired spin with two ^{27}Al nuclei, in agreement with Dzhabiev *et al.* [27], as well as a signal which could not be positively assigned.

Hiraki *et al.* [28] made ESR measurements during the polymerization of methyl methacrylate with catalyst systems based on $Ti(OC_4H_9)_4 + Al(C_2H_5)_3$ at Al/Ti molar ratios ranging from 2 to 15. They found that partially alkylated Ti^{3+} ions are active in homogeneous polymerizations in agreement with other studies on similar systems [29–31].

ESR studies on the $TiCl_4 + (C_2H_5)_2AlCl$ system indicated four types of resonance signals [32]. The presence of type I titanium ions, Ti^{3+}, was observed when the components were mixed at −70°C. On increasing the temperature, resonance signals with a hyperfine structure corresponding to ^{47}Ti and ^{49}Ti isotopes (with nuclear spins of 7/2 and 5/2, respectively) (type II) are observed. Type IV signal is a broadened asymmetric line. At −20°C, a signal with a supplementary hyperfine structure provided by the Al isotope (type III) appears. The authors analysed the ESR signal intensities as a function of reaction temperature. Ions of type I and II are found in the liquid phase. ESR spectra of Ti^{3+} ions (type III) are similar to the spectrum of crystalline β-$TiCl_3$, while those of type IV Ti^{3+} ions are similar to the ones corresponding to the α- and γ-$TiCl_3$ modifications [33]. The resulting complex takes the following structure:

$$TiCl_4 + (C_2H_5)_2AlCl \longrightarrow Cl_3Ti\begin{matrix} Cl \\ \\ Cl \end{matrix}Al\begin{matrix} C_2H_5 \\ \\ C_2H_5 \end{matrix} \quad (7)$$

which is the basic compound in the formation of Ti^{3+} ions. It was assumed that alkyl derivatives of Ti^{4+} formed by exchange of chlorine ligands with alkyl groups at Ti and Al atoms are present in the system. As a result of the high lability of the Ti–ethyl bond, such compounds are easily dissociated, resulting in the formation of Ti^{3+} ions.

ESR studies on Ziegler–Natta-type catalyst systems revealed that the *g*-values change with the degree of alkylation of the titanium. As long as $TiCl_4$ participates in the reaction with the aluminium alkyl resulting in the formation of a new alkylated product, *g*-values ranging between 1.92 and 1.93 are observed [21, 22]. If the alkylated product formed in this reaction is $TiCl_3R$, the signals have higher *g*-values, namely between 1.97 and 1.98 [17, 18, 23]. The $3d^1$-electron of the Ti^{3+} ion in such inorganic complexes gives a resonance signal with an anisotropic *g*-factor taking values between 1.91 and 1.93. Such data are important for the identification of discrete phases present in the catalyst system.

(*b*) *Pre-formed TIBA + $TiCl_4$ catalyst system*

Electron spin resonance spectra were recorded with a JEM-MS-3X spectrometer operating in the X-band (8.9–9.5 GHz). Spectrometer sensitivity is about 10^{11} spins/g for a temperature interval from −140° to +200°C.

Samples with Al/Ti molar ratios of 0.60, 0.75, 1.00, 1.25 and 1.50 and various maturation times were analysed. The samples, consisting of a reddish-brown slurry in the pre-forming medium, were introduced into quartz tubes under an inert-gas (argon), thus preventing the degradation of the catalyst complex. The resonance spectra of some samples recorded at room temperature are shown in Figures 7a–d.

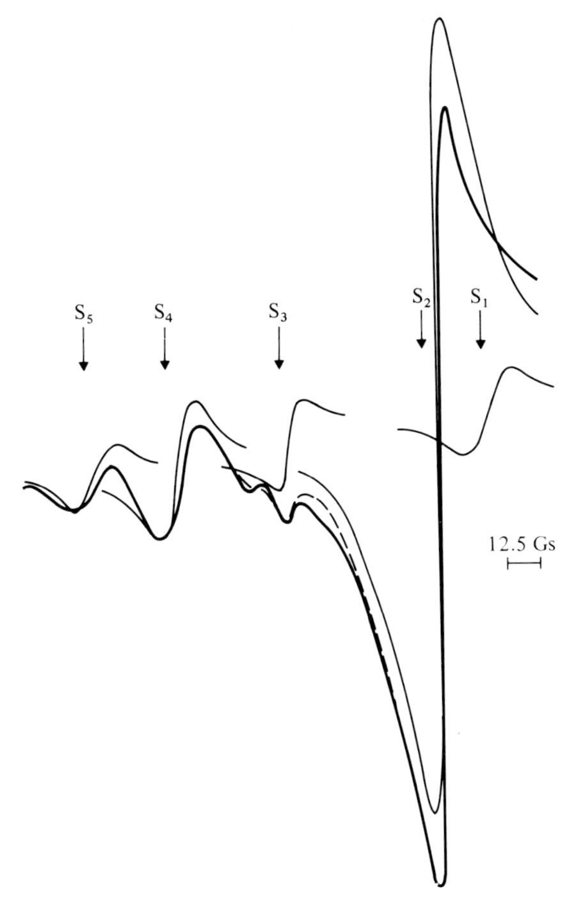

FIGURE 7. ESR spectra of the TIBA-$TiCl_4$ catalytic complex at room temperature; Al/Ti molar ratio: a = 0.6; b = 1.0; c = 1.25; d = 1.5. (——) spectrum components; (- - -) calculated spectrum.

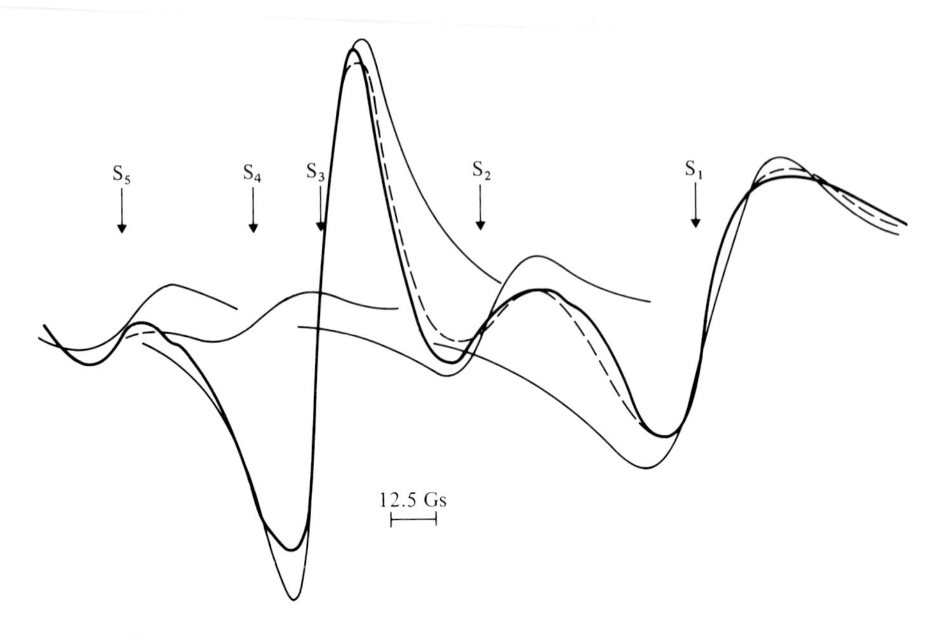

FIGURE 7b.

ESR spectra are complex, formed by the overlapping of at least five resonance lines. In all the samples, a resonance line near $g = 2.00$ was observed, which is attributed to the presence in the system of a free radical which is probably the isobutyl radical.

The maximum total number of paramagnetic centres (Ti^{3+}) was observed in catalyst complexes prepared at the optimum Al/Ti molar ratio of 1, as shown in Figure 8. The number of paramagnetic centres (Figures 8, 9, 11a and 11b) was determined with a sample of diphenylpicryl hydrazyl.

The dependence of the total number of paramagnetic centres on the maturation time of the catalyst complex at various Al/Ti molar ratios is illustrated in Figure 9. The number of paramagnetic centres (Figures 8, 9, 11a, 11b) was determined with a calibration sample of diphenylpicryl hydrazyl.

The number of paramagnetic centres was initially high and increased insignificantly, reaching a maximum after 8 to 10 hours. This is the time needed for the completion of the reduction reaction of $TiCl_4$ by TIBA. After this interval of time, the number of paramagnetic centres slowly decreases for up to about 24 hours, and then drops sharply.

The complexity of ESR spectra is also emphasized by the dependence of the signal intensity on the microwave power. In all cases, signal intensity is not linearly correlated with the reciprocal power, which is characteristic for spectra consisting of several lines and where the line saturation occurs at different values of the microwave field.

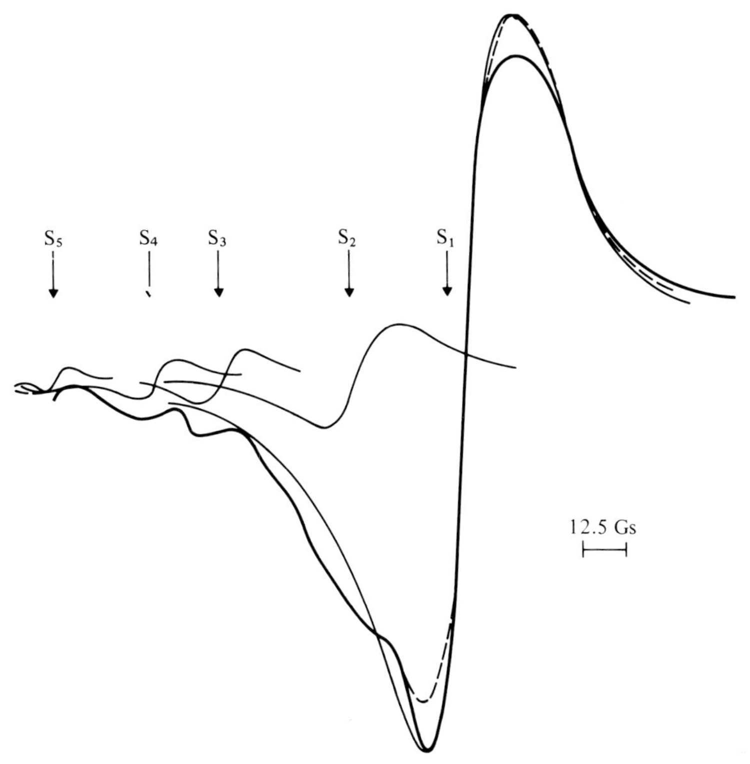

FIGURE 7c.

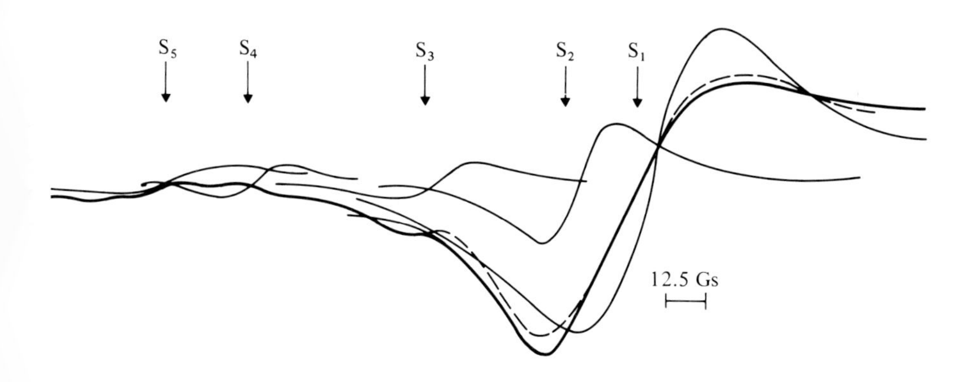

FIGURE 7d.

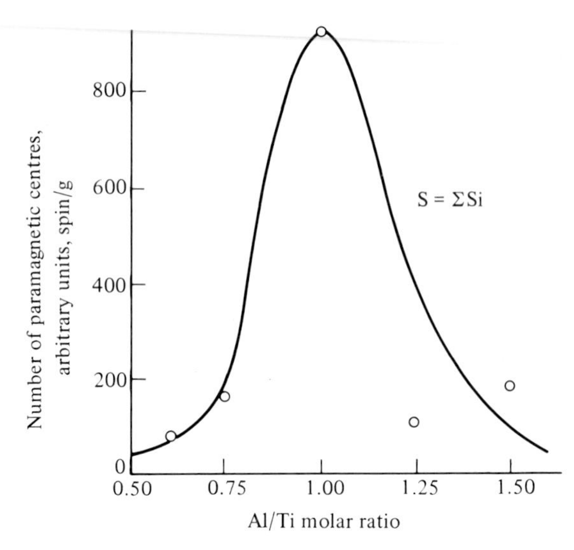

FIGURE 8. Variation of the total number of paramagnetic centres (Ti^{3+}) vs. the Al/Ti molar ratio.

Computer analysis of experimental spectra provided identification of various paramagnetic centres. Experimental data showed that as a function of the reaction conditions of $TiCl_4$ with TIBA in toluene, a mixture of complexes containing Ti^{3+} paramagnetic centres is obtained. The equilibrium composition of the system can be characterized by the relative intensity of the single lines.

The characteristic g-values of the resonance lines of the spectra are given in Figure 10. Such values permit the identification of complex components carrying Ti^{3+} ions in agreement with the analysis made in Section 1.6.1 on the correlation of g-values with the degree of alkylation of the titanium.

Five types of paramagnetic Ti^{3+} centres can be identified in the catalyst system according to the g-values. The resonance line, S_1, with g-value = 1.93 can be assigned to a non-alkylated form [21, 22]. By replacing the chlorine atoms attached to Ti^{3+} by alkyl groups, resonance lines with higher spectroscopic splitting factors are observed.

The alkylated forms of the $TiCl_3R$ type complexes which have g-values between 1.97 and 1.98 can be identified by the S_4 and S_5 components of the analysed system. The S_2 and S_3 lines can be assigned to some components with intermediate degrees of alkylation [17, 18, 23].

The correlation between the relative content, S, of paramagnetic centres in the S_1–S_5 components and the Al/Ti molar ratio is illustrated in Figure 11 (see also Figure 10).

For each single component of the complex, apart from S_2, a marked maximum is observed at an Al/Ti molar ratio of 1.0. The dependence of the paramagnetic centre number contributing to the S_2 line is not strictly correlated with the composition of the catalyst system.

The variation patterns of the total number of paramagnetic centres, S, as well as their relative content of S_1, S_3, S_4 and S_5 components, with the Al/Ti molar ratio and the maturation time are similar to the dependence of the catalytic activity of the system in isoprene polymerization on the Al/Ti molar ratio and the maturation time, as shown in Figures 3 and 5.

This fact emphasizes the close correlation between the catalyst activity and the number of Ti^{3+} ions present in the system and proves the importance of the ESR method in evaluating the efficiency of Ziegler–Natta-type catalyst complexes.

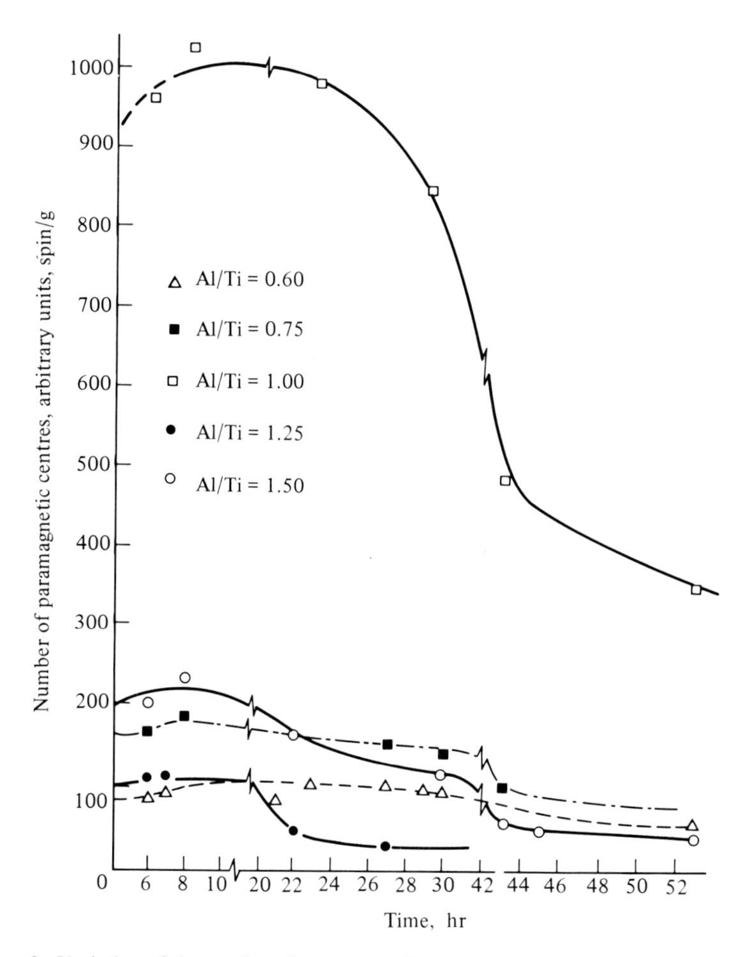

FIGURE 9. Variation of the number of paramagnetic centres vs. the maturation time, at various Al/Ti molar ratios.

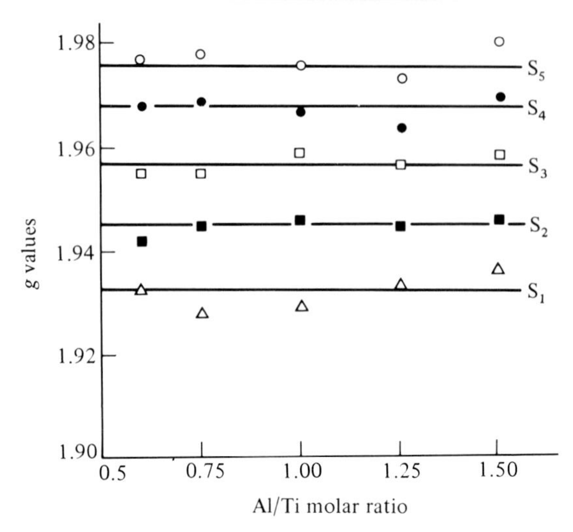

FIGURE 10. g-values of the resonance lines in ESR spectra of the TIBA + $TiCl_4$ catalyst system at various Al/Ti molar ratios.

2. Modified AlR_3 + $TiCl_4$ catalyst systems

The use of bicomponent Ziegler–Natta-type catalyst systems, even if preformed, results in a limitation of the polymerization reaction rate and of the resulting polyisoprene characteristics. The addition of electron donor

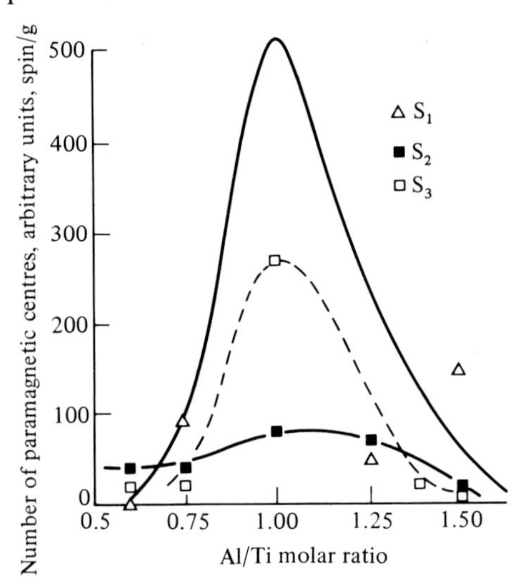

FIGURE 11a. Variation of the relative content of paramagnetic centres in S_1–S_3 components as a function of Al/Ti molar ratio.

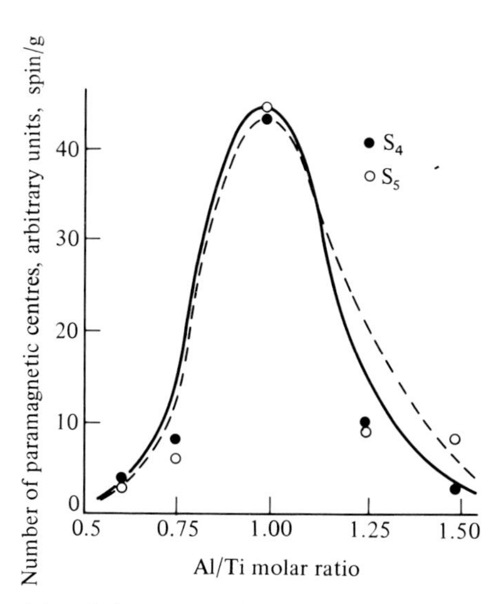

FIGURE 11b. Variation of the relative content of paramagnetic centres in S_4 and S_5 components as a function of Al/Ti molar ratio.

compounds (Lewis bases), i.e. amines [34–40], ethers [13, 34, 41–46], thioethers [47], alcohols [48], phenols [49], in certain ratios, either to one of the catalyst components or to the catalyst complex itself, affects the activity of the catalyst complex and, by implication, the course of the polymerization reaction, as well as the resulting polymer characteristics. Such compounds are called *modifiers*, while the catalyst complexes containing them have been termed "modified" Ziegler–Natta catalysts.

All the above observations, concerning the influence of the investigated factors on the activity of pre-formed bicomponent Ziegler–Natta-type catalyst complex, are still valid for the modified Ziegler–Natta catalyst complexes too, since at least they also are pre-formed and the basic components remain the same: TIBA and $TiCl_4$.

The use of modified Ziegler–Natta catalyst complexes, $AlR_3 + TiCl_4$, in the polymerization of isoprene shows certain advantages as compared with the two-component ones, namely:

— an increased polymerization reaction rate,
— a decrease in the polymer gel content,
— a shift toward higher levels of conversion when the properties of the polyisoprene are optimum.

It was also noticed that the presence of electron donors had no effect on the microstructure of the polymer.

The various assumptions on the mode of action of the modifiers in Ziegler–Natta-type systems were not clearly confirmed. However, it is obvious that the activity of such systems changes as a function of the qualitative and quantitative composition of atoms and groups within the coordination sphere of the transition metal, titanium in this case [34, 50, 51]. Several approaches were suggested on the mode of action of modifiers in stereospecific polymerization:

(a) the modifier blocks a great number of vacant neighbouring coordination sites and reduces the probability of the termination reaction [34];
(b) the donor forms complexes with the aluminium alkyl and lowers its net concentration; the rate of the chain-transfer reaction to the aluminium alkyl, present in the system, is consequently slowed down [52];
(c) the modifiers form complexes and, hence, they activate a "poisonous" metal alkyl ($AlRCl_2$) formed in the reaction between the catalyst components [38];
(d) the electron donors change the energy state of the surface of the catalyst complex and activate even the less active sites [53, 54];
(e) the electron donors facilitate a shift along the cleavage plane by interaction with the $TiCl_3$ crystal surface, which produces an increased number of active sites [55];
(f) the electron donors increase the number of defects on the crystal surface where polymerization takes place [41];
(g) the electron donors increase the alkyl unimeric fraction by complex formation with aluminium alkyl resulting in an increased capacity to form active sites [56–59].

Most of the preceding assumptions do not take into account the actual complex interactions occurring in the catalyst system. The observed donor effect is a result of various simultaneous interactions which depend upon the structure of the metal alkyl, the compound formed with the transition metal and the electron donor itself.

2.1. Addition of modifiers to aluminium alkyls

The agreement between hypotheses and experimental data is also verified by the occurrence of an optimum modifier/aluminium alkyl ratio giving maximum catalytic activity by the addition of electron donors to aluminium alkyls. This optimal ratio largely depends on the type of complex formed by the electron donor with the aluminium alkyl, and it is usually identical with the composition of the aluminium alkyl–electron donor complex; for instance, if the resulted complex is 1:1, the optimum electron donor/aluminium alkyl ratio is also 1/1. When two complexes can be formed, in 1:1 and 2:1 ratios, two maxima of catalytic activity were observed, which correspond to the interaction maxima [60].

Such optimal ratios are determined by chemical and physico-chemical methods.

The chemical method which consists of several polymerizations at various electron donor/aluminium alkyl ratios is tedious, discontinuous and time-consuming.

Among the physico-chemical methods, electrical conductivity measurements determine the optimum electron donor/aluminium alkyl ratio. The method needs a relatively simple equipment, is rapid and continuous [23, 58–61].

This method was applied to study such systems. Those electron donors were chosen which finally result in Ziegler–Natta-type catalyst complexes active in the stereospecific polymerization of isoprene and in the synthesis of polymers with improved properties [55]. As is well known, aluminium alkyls are usually in a dimeric form, owing to their electron deficiency (Lewis acids) which is cancelled by the formation of organic (R) bridges between the aluminium atoms [23, 58–65].

An equilibrium is established between dimeric and unimeric forms of aluminium alkyl, according to the equation [63, 64]:

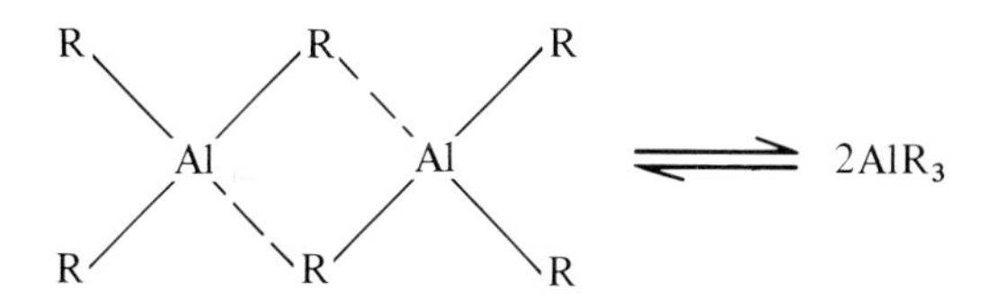

This equilibrium is more or less shifted to the right as a function of the length of the organic moiety (R), its bulkiness, temperature and concentration. The longer and bulkier the organic moiety and the higher the temperature and the dilution, the more is the equilibrium shifted towards the formation of monomeric aluminium alkyl.

Upon addition of electron donor compounds into solutions of aluminium alkyls in hydrocarbon solvents, the formation of complexes of Lewis base–Lewis acid type between the unimeric aluminium alkyl and the electron donor takes place, shifting the equilibrium towards the right:

$$\begin{aligned}(AlR_3)_2 &\rightleftarrows 2\,AlR_3 \\ AlR_3 + L &\rightleftarrows R_3Al \leftarrow L \qquad (L = \text{electron donor}).\end{aligned} \tag{9}$$

The purpose is completely to convert all aluminium alkyl into the unimeric form complexed with the electron donor. This corresponds to the optimal electron donor/aluminium alkyl ratio for a Ziegler–Natta catalyst complex with a maximum catalytic activity.

Based on the fact that unimeric aluminium alkyls have a higher electrical conductivity than dimeric aluminium alkyls and that the addition of electron donors results in an increased conductivity, it was assumed that complete dissociation and complexing of the aluminium alkyls correspond to a maximum electrical conductivity in the system which is further related to the maximum catalytic activity of the Ziegler–Natta-type complex formed.

The electron donor/aluminium alkyl molar ratio which leads to maximum electrical conductivity and therefore to a complete complex formation between the two reagents, depends on the steric hindrance of both components and on the number of donor groups present in the electron donor molecule. In the case of low or moderate steric hindrance and of only one donor group in the electron donor molecule, the maximum electrical conductivity should occur at a molar ratio equal to unity or approaching this value. The ratio increases with increased steric hindrance, provided no other factors interfere.

We studied two aluminium alkyl-electron donor systems, i.e. triisobutylaluminium-anisole and triisobutylaluminium-diphenyl ether, since such compounds are most interesting in the stereospecific polymerization of isoprene. Toluene was chosen as a solvent with a low dielectric constant ($\varepsilon_{20^\circ} = 2.36$), usually used in pre-forming Ziegler–Natta-type catalyst complexes.

The data which were obtained were correlated with the catalytic activities of heterogeneous Ziegler–Natta-type catalyst complexes formed from the above-mentioned etherates of triisobutylaluminium and titanium tetrachloride at an optimum Al/Ti molar ratio for isoprene polymerization.

The complexing reaction between TIBA and the electron donors studied is as follows:

$$(i\text{-}C_4H_9)_3Al + :\underset{\displaystyle R}{\overset{\displaystyle C_6H_5}{\overset{|}{\underset{|}{O}}}}: \rightleftarrows (i\text{-}C_4H_9)_3Al \leftarrow \underset{\displaystyle R}{\overset{\displaystyle C_6H_5}{\overset{|}{\underset{|}{O}}}}: \qquad (10)$$

where R = CH_3 for anisole and C_6H_5 for diphenyl ether.

Complexes of the 1:1 type are formed with both electron donors, but maximum electrical conductivities were recorded at different molar ratios, probably because of the difference in steric hindrance of the two electron donors.

(*a*) *Triisobutylaluminium–anisole system*

This system was studied in the temperature interval from $-10°$ to $+30°C$ at an original concentration of 100 mmoles/litre of TIBA. As shown in Figure 12,

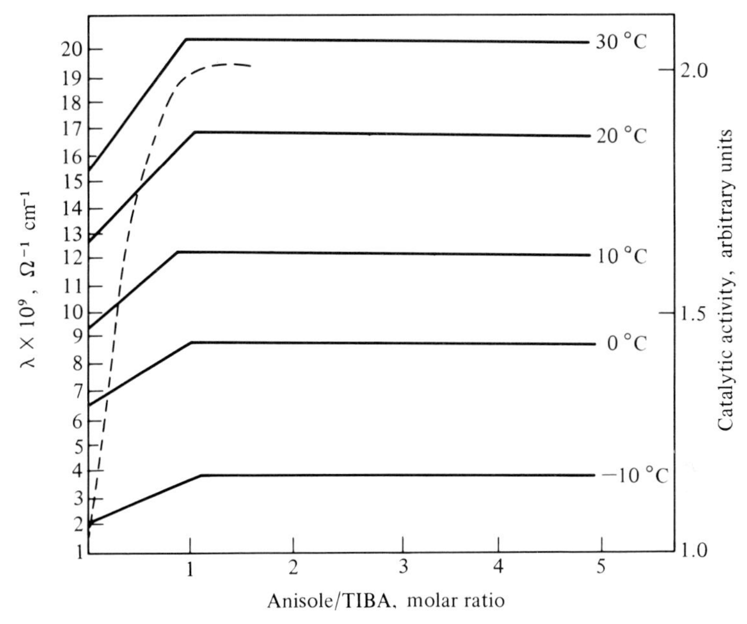

FIGURE 12. Electrical conductivity of TIBA–anisole system as a function of the molar ratio of the components and of temperature. The dashed curve represents the activity in isoprene polymerization of the Ziegler–Natta-type catalyst complex prepared from TIBA–anisole and $TiCl_4$.

the electrical conductivity increases rapidly up to an anisole/TIBA molar ratio of unity at all temperatures and further remains constant. The formation of the 1:1 complex occurs even at an equimolar ratio of the components. The rates of isoprene polymerization with Ziegler–Natta-type catalyst complexes prepared with such systems follow an ascending curve similar to the electrical conductivities, depending on the same variable, i.e. the electron donor/TIBA molar ratio. The maximum electrical conductivity is in agreement with data in the literature reporting a maximum catalytic activity for Ziegler–Natta-type catalyst complexes prepared with such systems at anisole/TIBA molar ratios of unity and above [41].

The corresponding electrical conductivity at a given ratio was reached almost instantaneously after the addition of the electron donor (anisole). This indicates a high rate of complex formation. The fact that the highest electrical conductivity is obtained at a molar ratio of unity at all temperatures means that in all cases the almost total complex formation of TIBA takes place at this ratio.

Since the maximum conductivity is clearly obtained at an equimolar ratio of the components, anisole can also be used for the determination of TIBA by the conductometric method in hydrocarbon solvents.

(b) Triisobutylaluminium–diphenyl ether (DPE) system

This system too was studied in the same temperature range (−10° to +30°C), as the previous one, at the same initial concentration of 100 mmoles/litre TIBA.

The maximum electrical conductivity is reached at a DPE/TIBA molar ratio ranging between 1.6 and 2, as a function of temperature, as shown in Figure 13. The actual value of this maximum is not as precise as for the TIBA–anisole system. At a temperature of +30°C, the maximum electrical conductivity is reached at a DPE/TIBA molar ratio of about 1.6, while at −10°C, the ratio is 2. It was proved that both the DPE–TIBA complex and the anisole complex are of the 1:1 type [66]. However, steric hindrance probably occurs in the first case since the methyl group was replaced by the bulkier phenyl group, showing the lower complexing ability of diphenyl ether as compared to anisole.

The differences of DPE/TIBA ratios corresponding to maximum electrical

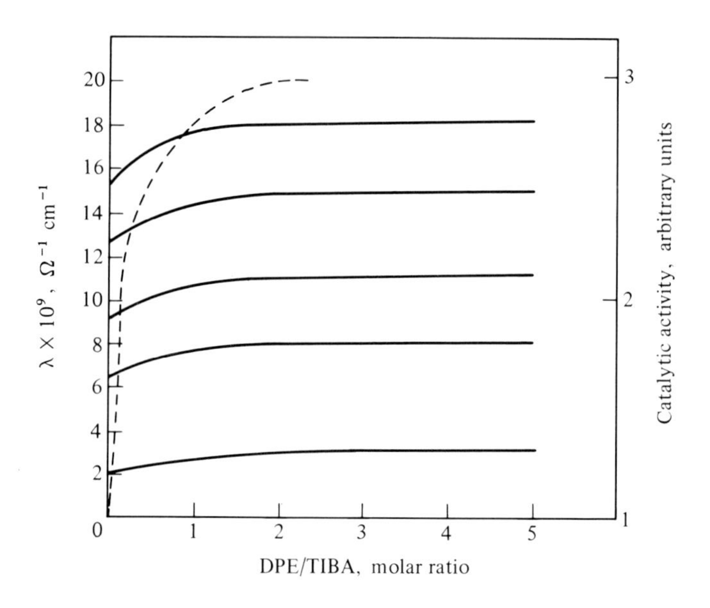

FIGURE 13. Electrical conductivity of TIBA–DPE system as a function of the molar ratio of the components. The dashed line represents the activity in isoprene polymerization of the Ziegler–Natta-type catalyst complex pre-formed from TIBA–ether and $TiCl_4$.

conductivity at various temperatures could be explained by the diminishing ability of diphenyl ether to form complexes as the temperature is decreased. These data too are in good agreement with those related to the catalytic activity of Ziegler–Natta complexes which reach their maximum values at a DPE/TIBA molar ratio of *ca.* 2. For economic reasons, a DPE/TIBA molar ratio $\leqslant 1$ is used in most manufacturing processes for *cis*-1,4-polyisoprene rubber. It yielded better results than the bi-component systems and is technologically satisfactory [41–44].

The complex formation mechanism is assumed to be identical in both systems: the electron donor, i.e. the aromatic ether, forms a complex with monomeric TIBA, shifting the monomer–dimer equilibrium towards the monomeric form as far as a total dissociation of the dimeric form [58, 59]. The complete formation of a complex of the TIBA in its monomeric form with the electron donors is highly important, as these complexes are extremely reactive with the other catalyst component, $TiCl_4$. The importance of complex formation is more obvious, if one keeps in mind that low temperatures favour an equilibrium shift towards the dimeric form and that catalyst complexes are usually prepared at such temperatures (usually below $-20°C$).

A relatively great number of modified Ziegler–Natta catalyst complexes, where the modifier was added to the aluminium alkyl (TIBA) solution, were studied. Among all the modifiers examined, i.e. triphenylamine, tri-*n*-butylamine, *n*-butyl ether, isobutyl ether, isoamyl ether, anisole, and diphenyl ether, the best results were obtained with the last one.

Experimental results on the polymerization of isoprene with Ziegler–Natta-type catalyst complexes based on $TiCl_4$ and TIBA complexed with diphenyl ether, pre-formed in toluene, are given in Table 5.

From the data shown in Table 5, we concluded that the optimal DPE/TIBA molar ratio, within the range studied, is about 1–2. In this interval, under the same pre-forming conditions, no significant differences in catalyst activity are noticeable, in spite of the DPE/TIBA molar ratios used. At DPE/TIBA molar ratios lower than unity, the catalyst complexes show a lower activity, although they are higher than the unmodified ones and are satifactory for technological purposes. These results are in good agreement with the above-mentioned data recorded by electrical conductivity measurements on the optimum DPE/TIBA molar ratio.

A comparison of the data listed in Tables 5 and 3 suggests that better results are obtained with Ziegler–Natta-type catalyst complexes modified with diphenyl ether than with the pre-formed two-component system.

The rate of polymerization of isoprene with the modified catalyst complex is faster than if a two-component catalyst complex is used. In the pre-forming temperature interval from $-40°$ to $-20°C$, the concentration of the catalyst could thus be reduced to half its value, without affecting the polymerization rate (especially at DPE/TIBA molar ratios of unity and above).

Table 5

Effect of DPE/TIBA molar ratio and of preformation temperature of the catalytic complex on the activity of the catalytic system $TiCl_4$ – (DPE-TIBA) in isoprene polymerization

	Catalyst			Polyisoprene characteristics		
No.	DPE/TIBA, molar ratio	Preforming temperature, °C	Conversion, %	$[\eta]$[a] dl/g	Gel, %	Tacticity, *cis*-1,4 %
1	2	−40	99	5.43	16	96.5
2		−30	94	5.38	15	96.2
3		−20	89	5.41	16	95.9
4		−10	84	5.30	17	95.9
5		0	67	5.36	17	95.6
6		10	44	5.28	16	96.9
7		20	30	5.30	18	96.3
8		30	23	5.35	16	96.5
9	1.5	−40	98	5.38	16	94.8
10		−30	95	5.36	20	95.9
11		−20	87	5.22	16	96.2
12		−10	86	5.31	16	96.1
13		0	65	5.31	16	95.7
14		10	41	5.29	18	94.9
15		20	29	5.29	16	95.6
16		30	24	5.38	17	96.0
17	1.0	−40	95	5.33	16	94.8
18		−30	92	5.21	15	96.3
19		−20	87	5.37	17	96.9
20		−10	79	5.41	18	95.7
21		0	59	5.37	16	95.7
22		10	38	5.29	16	95.9
23		20	27	5.41	17	96.1
24		30	24	5.40	15	96.0
25	0.5	−40	81	4.97	17	94.8
26		−30	78	4.98	17	95.9
27		−20	74	5.06	16	96.4
28		−10	69	4.97	16	96.3
29		0	47	5.00	16	95.5
30		10	22	5.01	18	94.6
31		20	14	4.95	19	95.7
32		30	12	4.91	16	96.1

[a] In toluene at 30°C.

The catalytic complex performed in toluene and maturated for 1 hr at room temperature; Al/Ti molar ratio = 1; $TiCl_4$ concentration = 0.3 g/100 g isoprene; polymerization time = 2.5 hr.

The average molecular weight, as determined by the intrinsic viscosity, shows a significant increase when modified catalyst complexes are used, as compared with the unmodified ones, especially at DPE/TIBA molar ratios of unity and above.

2.2. Addition of modifiers to $TiCl_4$

Another way to prepare a modified Ziegler–Natta-type catalyst complex for the polymerization of isoprene involves addition of the modifier to the $TiCl_4$ solution which is then mixed with the TIBA solution. Such a catalyst, where diphenyl ether was added to the $TiCl_4$ solution at DPE/$TiCl_4$ molar ratios higher than one, is reported in the literature. After the reduction reaction with TIBA, a catalyst complex results which is highly active in the stereospecific polymerization of isoprene [44].

It is difficult to state the real mode of action of the modifier in this case. $TiCl_4$ probably forms complexes with the electron donor in the same way that triisobutylaluminium does, since titanium contains vacant orbitals. Our attempts to identify, by electrical conductivity measurements, complexes of the same type as those formed between TIBA and anisole or diphenyl ether, respectively, have failed. The explanation could be that $TiCl_4$ is a weaker Lewis acid than TIBA and the complexes formed with the above-mentioned ethers are unstable at the relatively high temperatures (over $-40°C$) used in experiments. Therefore, the equilibrium is strongly shifted towards the individual components (dissociation of the complex) [15].

Isolation of some complexes of $TiCl_4$ with dioxane, tetrahydrofuran, anisole, and diisopropyl ether is reported in the literature, but the experiments were made at much lower temperatures (below $-150°C$). Complex formation occurs via the unshared electrons of the ether oxygen and the vacant hybrid d^2sp^3-orbitals of titanium [67]. Hence, the optimum modifier/$TiCl_4$ ratio could not be established by the electrical conductivity technique.

Starting from the assumption that diphenyl ether when complexed with $TiCl_4$ leads to highly active Ziegler–Natta-type catalysts and that aromatic hydrocarbons, present in the catalyst system, also improve catalytic activity, we studied, as a modifier added to $TiCl_4$, the eutectic mixture containing 71 moles % of diphenyl ether and 29 moles % of diphenyl. This mixture, known under the trade names of Diphyl or Dowtherm, is a heat-transfer agent frequently used in the chemical industry. Diphyl has the advantage of being liquid at room temperature and therefore can be more easily handled and metered [14].

In this case too, as when the modifier was added to triisobutylaluminium, the influence of the pre-forming temperature in the $-45°$ to $0°C$ range and of the modifier/$TiCl_4$ ratio on the activity of the catalyst system prepared from

Table 6

Effect of catalyst preformation temperature and of the modifier/$TiCl_4$ molar ratio on the activity of the (Diphyl + $TiCl_4$) + TIBA catalytic system in isoprene polymerization

	Catalyst			Polyisoprene characteristics		
No.	Modifier/ $TiCl_4$ molar ratio[a]	Preforming temperature, °C	Conversion, %	$[\eta]$,[b] dl/g	Gel, %	Tacticity, *cis*-1,4 %
1	1.00	−45	99	5.42	21	96.7
2		−30	97	5.47	19	95.9
3		−20	93	5.61	20	95.9
4		−10	87	5.40	20	96.1
5		0	75	5.42	19	95.3
6	0.75	−45	98	5.72	19	96.3
7		−30	98	5.61	20	96.7
8		−20	90	5.60	18	95.1
9		−10	81	5.47	18	95.9
10		0	72	5.49	20	96.1
11	0.50	−45	95	5.71	18	96.3
12		−30	92	5.63	20	96.0
13		−20	89	5.67	21	95.8
14		−10	79	5.58	20	96.3
15		0	63	5.43	18	95.8
16	0.25	−45	84	5.53	19	94.9
17		−30	76	5.47	21	96.1
18		−20	72	5.42	18	95.3
19		−10	67	5.39	18	95.3
20		0	58	5.43	18	96.0
21	0.16	−45	62	5.38	18	95.8

[a] Calculated as DPE/$TiCl_4$.
[b] In toluene at 30°C.
The complex was maturated for 1 hr at room temperature; Al/Ti molar ratio = 1; catalyst concentration 0.3 g $TiCl_4$/100 g isoprene; polymerization time 2.5 hr.

the (diphenyl ether + diphenyl)–$TiCl_4$ complex and TIBA was studied. The results are listed in Table 6.

An inspection of the data given in Table 6 leads to the conclusion that a decreased temperature of pre-forming produces an increased polymerization rate similar to the behaviour of both the two-component catalyst complex and the (DPE + TIBA) + $TiCl_4$ catalyst complex.

A decrease in the modifier/$TiCl_4$ molar ratio from 1 to 0.5 produced no significant reduction of the reaction rate. At a modifier/$TiCl_4$ molar ratio of 0.25, the polymerization rates were still convenient. It is only at a modifier/$TiCl_4$ molar ratio of 0.16 that the activity of the catalyst complex decreased.

One can state that Ziegler–Natta-type catalyst complexes based on the diphyl + $TiCl_4$ complex and TIBA show an activity similar to that based on

$TiCl_4$ and the diphenyl ether + TIBA complex; moreover, it presents some additional technological advantages.

Neither the molecular weight, nor the microstructure of the polymer are markedly influenced by the modifier/$TiCl_4$ ratio.

Chapter II

Stereospecific polymerization of isoprene with pre-formed catalyst complexes

THE STEREOSPECIFIC polymerization of isoprene in solution is related to a great number of factors such as: the nature of the catalyst system, the nature and purity of the reaction medium, the technological conditions, etc. As the catalyst systems have been previously discussed, this chapter deals with some questions related to the polymerization reaction itself.

1. Purity of polymerization systems

One of the most important problems is how to ensure adequate purity conditions. The nature of the impurities found in a monomer largely depends on its manufacturing process. Several manufacturing processes of isoprene are known, each of them raising specific purification problems to obtain a monomer which can be used in stereospecific polymerizations [14, 68–70]. The need to achieve a high degree of purity in the system is proved by the fact that an impurity content which is higher than certain permitted limits affects not only the catalyst activity, the reaction mechanism and its kinetics, but also the properties of the resulting product [71, 72].

The composition of some isoprene samples produced by various processes is shown in Table 7. The various qualities of isoprene can result in comparable polymerization reactions, despite their different isoprene concentrations, provided that the impurities present do not influence the catalyst activity. Polymerization of the mixture resulted from the dehydrogenation of isopentane was therefore possible, although it contained isoprene (20–25%) and pentenes (2-methyl-1-butene, 2-methyl-2-butene and 3-methyl-1-butene) [73].

The strongest catalyst inhibitor of the stereospecific polymerization of isoprene is cyclopentadiene, which destroys the catalyst at a concentration of 1.5×10^{-3} mole/litre. During the storage of isoprene containing cyclopen-

Table 7

Chromatographic analysis of isoprene synthesized by various methods (% v/v)

No.	Compound	From isobutylene and formaldehyde	From the C_5 fraction of cracked gasoline	One-step dehydrogenation of *i*-pentane
1	Propane	0.00035	—	—
2	Propylene	0.00007	—	—
3	Isobutane	0.0001	—	—
4	*n*-Butane	0.00153	0.00026	—
5	1-Butene + isobutene	0.00072	0.00056	—
6	*trans*-2-Butene	0.00041	0.00025	—
7	*cis*-2-Butene	0.13340	0.00097	—
8	*i*-Pentane	0.03849	0.01831	0.00068
9	3-Methyl-1-butene	0.14382	—	—
10	*n*-Pentene	0.00048	0.00699	—
11	2-Methyl-1-butene	0.28403	0.23784	0.11345
12	*trans*-2-Pentene	0.01194	0.35067	0.06717
13	*cis*-2-Pentene	0.02592	0.36781	0.03179
14	2-Methyl-2-pentene	0.62097	0.00030	0.77138
15	Isoprene	98.85625	98.84482	99.71219
16	Acetylenes (2-butyne)	0.00084	0.00029	—
17	Cyclopentadiene	0.00002	0.00011	0.00034
18	*n*-Pentane	—	0.15131	0.00272
19	1,3-Butadiene	—	0.06626	—
20	3-Methyl pentane	—	0.00009	—
21	Cyclopentane	—	0.00017	—
22	*trans*-Piperylene	—	0.00008	—
23	*cis*-Piperylene	—	0.00036	—
24	Pentynes	—	—	0.00028
25	x	—	—	0.00013

tadiene, the latter undergoes dimerization and the resulting dicyclopentadiene reacts, in its turn, with the active polymerization sites of the catalyst system [74]. The permitted maximum concentration of these two compounds in the monomer is only 5 ppm. This inhibiting property of cyclopentadiene enabled an indirect evaluation to be made of the active site concentration in the polymerization of isoprene with the TIBA + $TiCl_4$ catalyst system. The determination is based on the observation that an increased inhibitor concentration produces an initial sharp decrease in the polymerization reaction rate, which at still higher concentrations tends to level off. By extrapolation to zero rate, the number, and hence the concentration, of potentially active sites can be evaluated. In this way, the number of active sites in TIBA + $TiCl_4$ catalyst system could be estimated to 0.2–0.5% of the original titanium content [75, 76].

Other strong inhibitors of the catalyst system are dimethylformamide and butyl mercaptan which can enter the structure of the active sites and change the mechanism of diene addition. A special group of inhibitors of the catalyst system consists of acetylenic and allenic hydrocarbons which produce an increase in the induction period and a decrease in both the molecular weight and the polymerization rate. Indeed, from their effect on the reaction kinetics, the polar compounds which act as impurities can be classified into two groups: the first group includes those compounds which slow down the rate of polymerization without any effect on the induction period (diethyl sulphide, acetonitrile, diethylamine, carbon monoxide, diethyl ether, vinyl ethyl ether, water, thiophene, carbon disulphide, carbon dioxide, carbon oxysulphide). The second group, apart from a slowing down of the rate of polymerization, also produces an increase in the induction period (ethyl alcohol, acetone, methyl ethyl ketone, hydrogen sulphide, oxygen, formic acid, dimethylamine, ammonia) [74].

Laboratory experiments made in order to remove the most important inhibitors from isoprene showed that treatment with water and maleic anhydride in excess at a low temperature, followed by distillation on a rotating column of 100 theoretical plates and at a very high reflux ratio, ensures the removal of cyclopentadiene, dimethylformamide, dimethylamine and piperylene. In order to remove carbonyl compounds and acetylenic hydrocarbons down to a final level below 10 ppm, molecular sieves or a copper catalyst can be used [77, 78].

Industrial methods of purifying isoprene are usually based on distillation procedures using highly efficient columns.

2. Some aspects of isoprene polymerization

2.1. Influence of polymerization procedures

Solution polymerization was chosen for the polymerization of isoprene with Ziegler–Natta-type catalyst systems. This procedure has a relatively low thermal effect per volume unit of the reaction mixture, ensures a facile handling of the polymerization mass and simple metering.

Although the polymerization of isoprene in bulk is unlikely to become an industrial process, experiments on bulk polymerization of isoprene with the tercomponent catalyst system were undertaken (Table 8). The final conversion in polyisoprene and some of its properties shown in Table 8 led to the following conclusions:

— Bulk polymerization reactions usually occur under less controllable conditions than in solution, in agreement with the data reported in the literature [79]. This fact is illustrated by the molecular weights of the soluble fraction which are neither markedly different nor reproducible as a function of the concentration of the catalyst system.

Table 8
Bulk polymerization of isoprene with the (Diphyl + $TiCl_4$) + TIBA catalyst system

No.	Catalyst concentration mmoles $TiCl_4$/L isoprene	Polymerization time (hr)	Final conversion (%)	Polyisoprene characteristics	
				$[\eta]$[(a)] (dl/g)	Gel (%)
1	2.40	0.25	10.0	3.84	31.6
2	2.23	1.00	15.0	4.12	44.3
3	2.12	2.00	9.3	4.56	46.0
4	1.70	4.00	9.7	4.71	30.2
5	1.47	5.00	9.2	4.23	43.0
6	1.27	2.00	10.0	4.01	28.9
7	1.17	2.00	9.8	4.50	28.0
8	1.17	5.00	10.4	5.64	39.0
9	1.05	5.00	10.8	4.80	41.7

[(a)] In toluene at 25°C.
Al/Ti molar ratio = 1; Diphyl/titanium molar ratio = 1.

— The polymer gel content is very high. This may be due to the growing chain of the polymer being less readily desorbed from the active site on the catalyst surface and moving into the polymer solution [80–84]. This happens particularly at relatively high conversions, when the viscosity of the reaction mass is very high. Polymer chain immobilization on particles of the catalyst complex probably takes place, favouring their participation in cationic processes on the surface of heterogeneous catalyst complexes and causing an increase in the amount of the insoluble fraction.

— A conversion levelling at about 10–15% is obtained. Korotkov *et al.* [85] reported that in similar experiments, conversion reached 29% after 30 hours and *ca.* 55% after 170 hours. It is assumed that the active sites are irreversibly destroyed or covered with a thick polymer layer which prevents the diffusion of monomer towards the active sites.

A number of disadvantages are avoided if the polymerization is performed in solution. In such cases, the nature of the solvent, however, plays a major role because it affects both the kinetics of the polymerization reaction and the properties of the product. The influence of the solvent was studied using the following solvents as polymerization medium: butane, pentane, hexane, heptane, octane, isopentane, iso-octane and cyclohexane [86] (Figure 14).

It was found that under identical polymerization conditions, conversion decreases in the following order: (pentane, hexane, heptane) > (isopentane, butane) > (octane, iso-octane) > cyclohexane.

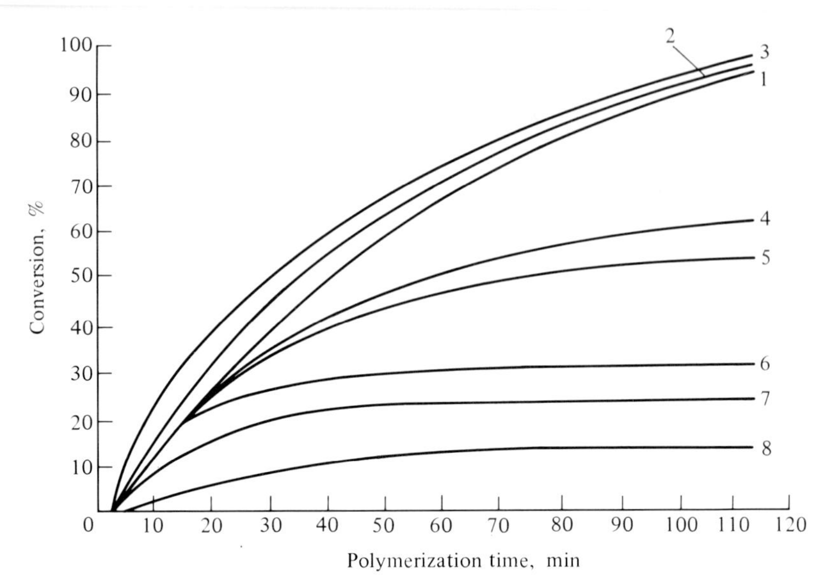

FIGURE 14. Effect of nature of solvent on conversion in isoprene polymerization. 1—pentane; 2—hexane; 3—heptane; 4—isopentane; 5—butane; 6—octane; 7—isooctane; 8—cyclohexane [86].

The rate of polymerization and the conversion depend on the solubility of the polymer in a given solvent. An increase in temperature at the start of the polymerization produces an acceleration of the rate of polymerization and sometimes an increased final conversion.

The molecular weight does not change up to a certain degree of conversion. Afterwards it can diminish during polymerization. This decrease in molecular weight depends on the nature of the solvents used. It was noticed that in isopentane, the polymer has the highest initial molecular weight, and therefore a relatively high final molecular weight.

No influence of the nature of the solvent on tacticity and physico-mechanical properties of pure gum vulcanizates was reported [86].

It is assumed that the rate of polymerization is directly proportional to the rate of polymer solubilization in the solvent. At a low rate of solubilization, the polymer covers the active sites of the catalyst which results in a decreased reaction rate and an increased gel content [87].

Polyisoprene shows the highest solubility in aromatic hydrocarbons (benzene, toluene). In such solvents, the reaction rate is high, while the gel content is low [84, 88, 89]. However, isopentane is the most widely used solvent on an industrial scale as the viscosities of the reaction mixture at convenient polymer concentrations are not too high and it has a low boiling point, which simplifies the further recovery process of the polymer.

Comparative polymerization runs carried out in heptane proved that the levels of conversion and the microstructure of the polyisoprene are in all cases virtually the same. No significant changes in molecular weight or in content of insoluble fraction occurred, so that similar physico-mechanical characteristics of the final product were obtained.

While individual hydrocarbons were used as solvents, experiments were simultaneously made with various petrochemical cuts.

The time of storage of the uninhibited monomer has a major influence on the course of the polymerization reaction. To determine the effect of this variable, we studied the variation with time of the temperature in polymerization experiments carried out under practically adiabatic conditions. As shown in Figure 15, the lengthening of the storage time produces an increase in the induction period. An explanation of this fact could be provided by the formation of isoprene dimers and trimers which affect the induction period of the polymerization [72].

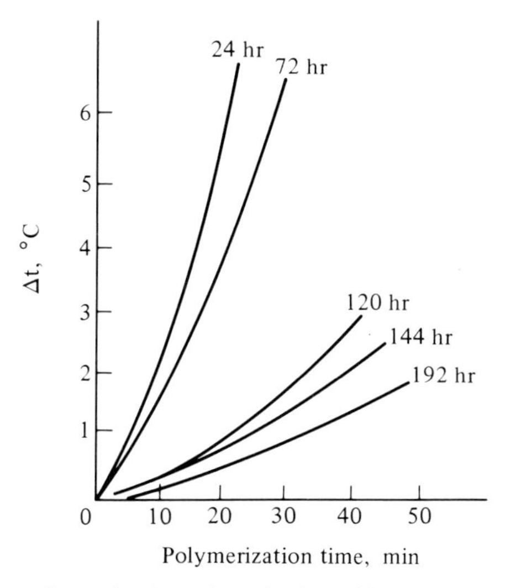

FIGURE 15. Reaction exotherms in the polymerization of isoprene which has been stored for various time intervals.

2.2. Viscosity of the polymerization mixture

In the solution polymerization of isoprene in the presence of stereospecific catalyst complexes, a solution is obtained the viscosity of which depends on the following factors: the content of polymer, the molecular weight of the polyisoprene, the temperature of the solution, the gel content of the polyisoprene and the nature of the solvent. The polyisoprenes prepared under similar polymerization conditions have comparable molecular weights. The viscosity of the solution increases during polymerization as a result of increased conversion. In polymerizations in a series of stirred reactors, the conversion and consequently the polymer concentration increase from each reactor to the next.

The variation of the viscosity of the polyisoprene solution as a function of polymer concentration, molecular weight, solution temperature and gel content is shown in Figures 16–19, respectively [90–92].

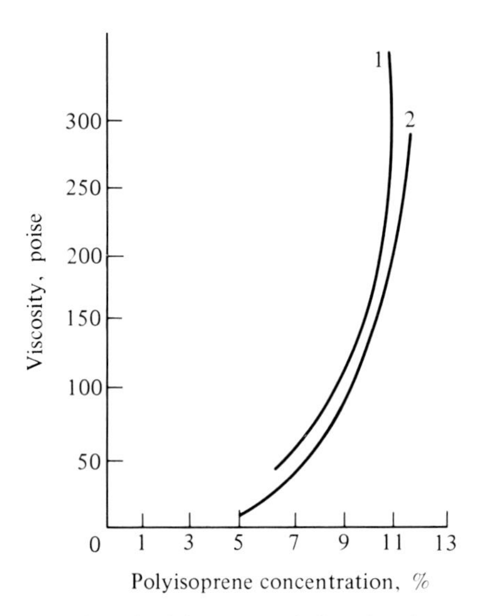

FIGURE 16. Variation of viscosity of polyisoprene solutions in toluene as a function of polymer concentration. Temperature: 20°C. $[\eta]$ = curve 1 = 3.2 dl/g; curve 2 = 3.1 dl/g [90].

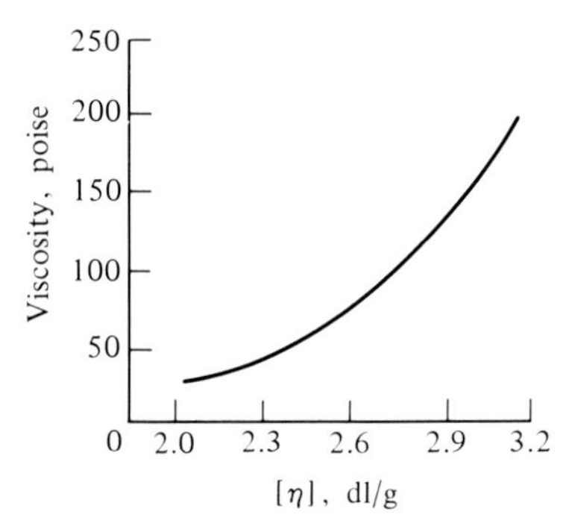

FIGURE 17. Variation of viscosity of polyisoprene solutions in toluene as a function of molecular weight. Polymer concentration = 10.5% by weight, solution temperature = 5°C [90].

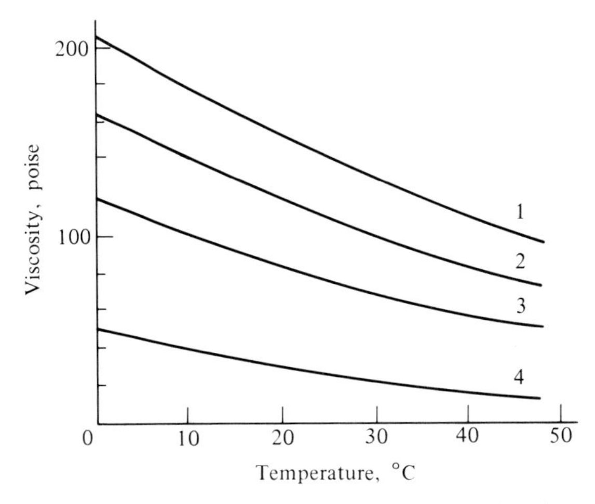

FIGURE 18. Dependence of the viscosity of polyisoprene solutions in toluene on temperature. Polyisoprene concentration = 10.5% by weight; [η]: curve 1 = 2.6 dl/g; curve 2 = 2.9 dl/g; curve 3 = 3.1 dl/g; curve 4 = 3.2 dl/g [90].

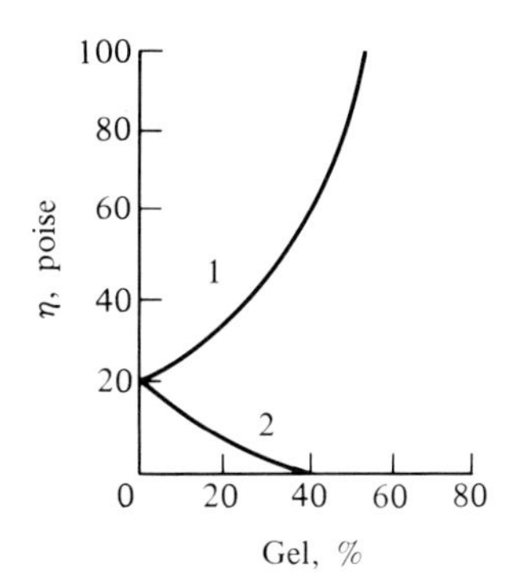

FIGURE 19. Dependence of the maximum Newtonian viscosity of solutions of polyisoprene on the gel content. Swelling index: curve 1, $\sigma = 30$; curve 2, $\sigma = 5$ [92].

The viscosity of solutions of polyisoprene in *n*-heptane changes considerably in the concentration range from 15 to 25% [91]. As shown in Figure 16, the viscosity of solutions of polyisoprene in toluene increases markedly at a much lower concentration, i.e. 8–10%, as for heptane solutions [90].

As shown in Figure 18, the viscosity of polyisoprene solutions decreases to about one-half of its value within the temperature range from 5 to 50°C. The viscosity of polymer solutions depends very much on solvent-polymer intractions, as indicated by data listed in Table 9 [90].

Table 9
Viscosity of polyisoprene solution as a function of the nature of the solvent [90]

Polyisoprene $[\eta]$[a] (dl/g)	Viscosity of the reaction mixture (poise)		
	Toluene	Heptane	Isopentane
3.2	205.6	23.3	16.4
2.3	78.3	11.4	8.4

[a] In benzene at 20°C.
Polymer concentration = 10.5%; temperature = 5°C.

The gel formed during the polymerization reaction has various effects on the viscosity of the reaction mixture, depending on its content and structure as expressed by the swelling index. The viscosity of a polyisoprene solution containing a dense gel (swelling index $\sigma < 15$) is considerably lower than a solution of the same concentration but with a loose gel (swelling index $\sigma > 15$) [92]. An increased content of loose gel produces a marked increase of the viscosity of the solution, as shown in Figure 19, while an increased content of dense gel results in a decrease in the viscosity of the solution, although a much greater number of molecules is included within the aggregate nucleus in the latter type of gel.

This can be attributed to a decrease in molecular polymer–solvent interactions leading to an apparent reduction of the concentration of the polyisoprene in the solution. Much less solvent can penetrate into the thick gel, owing to its agglomerated structure; the concentration of the sol fraction therefore is smaller, hence an overall decrease of solution viscosity occurs [92].

In determining the viscosity of the reaction mixture, we took into account the fact that the viscosity of a polymer solution varies with the shear stress. Solutions of macromolecular compounds have a non-Newtonian behaviour and their viscosity decreases with an increased shear stress. Hence, the viscosities of the reaction mixture resulting from the polymerization of isoprene can be expressed only as a function of the shear stress at which the measurements have been made.

The variation of the viscosity of a reaction mixture containing 13% of polyisporene before deactivation with the shear stress is shown in Figure 20, while the same variation is illustrated in Figure 21 for a 12% polyisoprene solution in *n*-heptane after deactivation of the catalyst complex.

The data shown in Figures 20 and 21 indicate that although the two polymer solutions are only slightly different in their content of polyisoprene, the viscosity of the polymer solution after deactivation is much lower than before this operation. Since this decrease in viscosity cannot be accounted for by changes in the thermodynamic properties of the polymer–solvent system,

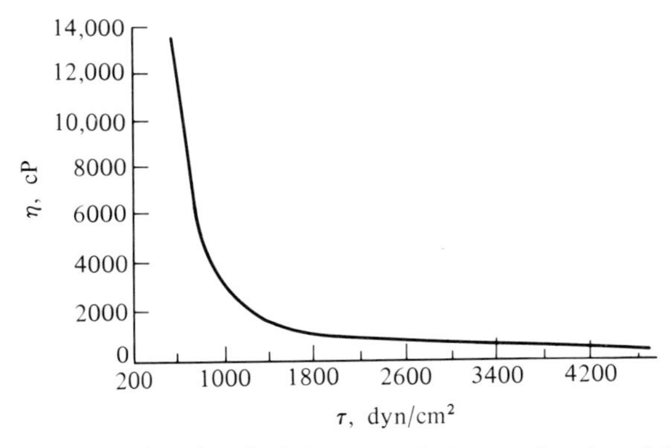

FIGURE 20. Variation of the viscosity of polyisoprene solution as a function of the shear stress before deactivation of the catalyst complex. Solvent: heptane; $[\eta] = 3.5$ dl/g; polyisoprene concentration = 13 %; temperature = 25°C.

as the amount of deactivating agent is below 1 % of the total volume of the system, it was assumed that the continuous polyisoprene network breaks down into fragments consisting of macromolecular aggregates or single macromolecules [81]. The unchanged part is found after separation and drying as the gel fraction of the polymer.

The high viscosity of polyisoprene solutions made it necessary that special polymerization reactors be designed fitted with new stirring equipment and heat exchangers. Polymerization reactors which ensure the stirring of the reaction mixtures with viscosities of *ca.* 1000 poises are now available [93].

In the continuous polymerization of isoprene with Ziegler–Natta-type catalyst systems, conversions of 30 to 50 % are usually reached in the first reactor. Due to the highly exothermic polymerization reaction, thermal shock in the first reactor is much higher than in the other reactors of the series.

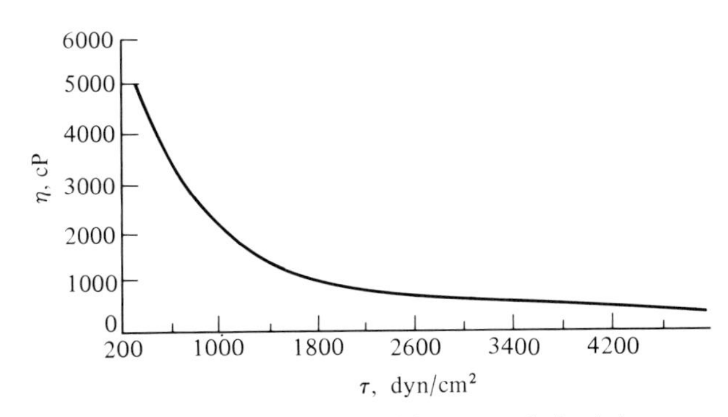

FIGURE 21. Variation of the viscosity of a 12 % polyisoprene solution in heptane as a function of the shear stress, after deactivation.

Therefore, attempts were made to level off the thermal effects of the reaction in two ways:

— the stepwise feeding of catalyst complex;
— the stepwise feeding of monomer.

The stepwise addition of the catalyst complex using the same total concentration results in lower final conversions which necessitate an increase in the unit consumption of catalyst components, unjustified from an economic point of view. No striking differences in the physico-chemical and physico-mechanical properties of the rubber are noticeable.

The kinetic equation of the overall reaction rate can be written as follows:

$$-\frac{d[M]}{dt} = k[M][C] \tag{1}$$

where [M] is the monomer concentration, [C] the catalyst concentration and k the rate constant of the polymerization reaction.

In the stepwise addition of monomer, the first reactor is fed with a monomer-solvent mixture which contains 5% of isoprene, while the rest of the isoprene (up to 12.5%) is introduced into the second reactor. By calculating the ratio between the polymerization rates of a mixture with an original isoprene content of 12.5% and of the stepwise reaction, it has been found that the thermal effect in the first reactor, and hence the conversion, are reduced by *ca.* 30%. The addition of isoprene into the second reactor increases its concentration and concomitantly the reaction rate which cancels the lower conversion in the first reactor and tends to equalize the thermal requirements of the two reactors. A more efficient levelling is attained by extending the above method to a greater number of the battery of reactors.

The molecular weight is not affected by the stepwise addition of monomer. From molecular weight distribution curves, determined by the ultracentrifuge [94], one observes a broadening due to the formation of larger amounts of polymer with high molecular weights.

2.3. Kinetics of the solution polymerization of isoprene

The interest in the automation of the polymerization process and the use of computers for this purpose stimulated more elaborate kinetic studies. The investigations were aimed at outlining the general character of the relationship between conversions on the one hand and catalyst and monomer concentrations and the temperature of the polymerization reaction on the other [95].

The mathematic equations better describing the polymerization course were derived by two basic approaches. One is based on the assumption that the polymerization process involves three main steps: initiation, propagation and termination; equations are derived based on theoretical hypotheses which are then experimentally tested. The process is arbitrarily divided into two steps as a function of the probability of the monomer molecule reaching an active site on the catalyst and subsequently being included in the growing chain of the polymer. During the first step, at conversions up to 25%, monomer diffusion towards the active site through the polymer layer surrounding the catalyst takes place quite easily and the reaction rate is expressed by the following equation [96, 97]:

$$-\frac{d[M]}{dt} \cong \alpha \frac{k_1 k_2}{k_3} [M]_0^2 \left([C]_0 - \frac{1}{K}\right) \qquad (2)$$

where k_1, k_2 and k_3 are the initiation, propagation and termination rate constants, respectively,

- α is a coefficient related to the state of dispersion of the catalyst,
- K is the association constant of an active catalyst site, according to the reaction: $TiCl_3 + R_2AlCl \rightleftarrows TiCl_3 \cdot R_2AlCl$,
- $[C]_0$ is the original catalyst concentration,
- $[M]$ is the monomer concentration at time t,
- and t is the polymerization time.

Integration of the above equation and comparison of the resulting expression with the available experimental data led to the following conclusions:

- the equation describes accurately the course of polymerization at conversions up to 25%;
- the overall polymerization rate is first order with respect to the concentration of the catalyst complex;
- the overall activation energy depends on the nature of the solvent in which the catalyst complex is pre-formed, e.g. 9.6 kcal/mole in benzene, and 14.9 kcal/mole in isopentane.

At conversions higher than 25%, the catalyst complex is covered with a layer of polymer which limits the diffusion of monomer towards the active sites and in such cases the process is described by the equation [96, 97]:

$$-\frac{d[M]}{dt} = \beta e^{-\pi/4} \left(\frac{D}{\pi}\right)^{1/2} [C]_0^{1/2} t^{-1/2} [M] \qquad (3)$$

where β is a proportionality coefficient relating the surface of the catalyst particle to its concentration, and D is the diffusion coefficient of the monomer through the polymer layer.

The above equation is difficult to use since it contains terms which cannot be directly measured. On the other hand, it is difficult to clearly delimit the range of validity of the two equations.

Therefore the method involving a general form of the equation of the polymerization rate is more widely used. In this case, the kinetic parameters included in the equation are established by statistical computation of a set of experimental data.

An equation of the following type can be written for the rate of isoprene polymerization [98–100]:

$$-\frac{d[M]}{dt} = k[M]^n \tag{4}$$

where M is the monomer concentration, and k is the reaction rate constant. If the catalyst is modified with a Lewis base (L), then k is a function of catalyst concentration, [C], modifier concentration, [L], reaction temperature, T, catalyst component Al/Ti molar ratio, r, and the temperature of pre-forming of the catalyst complex, T_{cat}.

Also, n, the reaction order with respect to monomer, is a function of catalyst concentration, [C], and modifier concentration, [L].

A general equation describing the development of the process, regardless of the actual values of the rate-determining variables has still not been deduced, but equations for certain particular cases, widely used in practice, have however been found. When the Al/Ti molar ratio of the catalyst components, $r = 1$, and the pre-forming temperature, $T_{cat} = -20°C$, equation (4) becomes

$$-\frac{d[M]}{dt} = k_1(T, [C], [L])[M]^{n([C],[L])} \tag{5}$$

where, according to Arrhenius law:

$$k_1(T, [C], [L]) = A([C], [L])\exp(-E([L])/RT) \tag{6}$$

where A is the pre-exponential factor

E, the activation energy,

R, the gas constant.

For the polymerization of isoprene with Al/Ti catalysts modified with diphenyl ether, computations proved that, regardless of the catalyst and the modifier concentrations, n is 0.52. The activation energy as well as the pre-exponential factor, A, depend on modifier concentration:

— at $[L] = 0.6$ mole/mole TIBA, $E = 7400$ cal/mole

$$A = (3.32[C] - 1.16 \times 10^{-3}) \times 10^7, \tag{7}$$

— at [L] = 0.9 mole/mole TIBA, $E = 5200$ cal/mole

$$A = (1.63\,[C] - 0.36 \times 10^{-3}) \times 10^{6}. \tag{8}$$

It will be observed that this kinetic treatment also confirms the first-order rate with respect to the concentration of the catalyst. The experimental observations on the reduction of the activation energy of the polymerization process by using catalysts modified with electron donors are also corroborated.

If one takes into account the influence of the Al/Ti molar ratio (r) and of the pre-forming temperature of the catalyst complex (T_{cat}), one can write [101]:

$$k = k_1 \left(1 - \frac{r-1}{A}\right)^{a} \left(1 - \frac{r-1}{B}\right)^{b} \tag{9}$$

where A, B, a and b are constants which depend on the pre-forming temperature of the catalyst complex:

— at $T_{cat} = -20°C$

$$k = 1.56 \left(1 - \frac{r-1}{0.021}\right)^{0.474} \left(1 - \frac{r-1}{0.440}\right)^{10}, \tag{10}$$

— at $T_{cat} = -40°C$

$$k = 1.62 \left(1 - \frac{r-1}{0.045}\right)^{1.932} \left(1 - \frac{r-1}{0.100}\right)^{4.295} \tag{11}$$

The above two kinetic equations offered theoretical explanations for two important experimental observations:

(a) the optimum pre-forming temperature of the catalyst complex is about $-20°C$, which is easily observed by plotting $\ln k = f(1/T_{cat})$ taking the optimal values of the molar ratio of the catalyst component (Figure 22) [101]. A sharp change of slope is noticed at a temperature of about $-20°C$, which is probably related to a modification of the ratios of the rates of the different reaction steps.

(b) A catalyst complex, kept up to 30 minutes, with an Al/Ti molar ratio $r = 1.05$ can be reactivated by the addition of either $TiCl_4$ or of a catalyst complex with $r = 0.95$. Reactivation is most likely if the complex is prepared and kept at lower temperatures [101, 102].

The course of a polymerization reaction carried out in a series of stirred reactors [100], the dynamics of an industrial polymerization process [103] and the overall kinetics of the process performed by stepwise addition of monomer [104] could be mathematically expressed by means of the above kinetic equations.

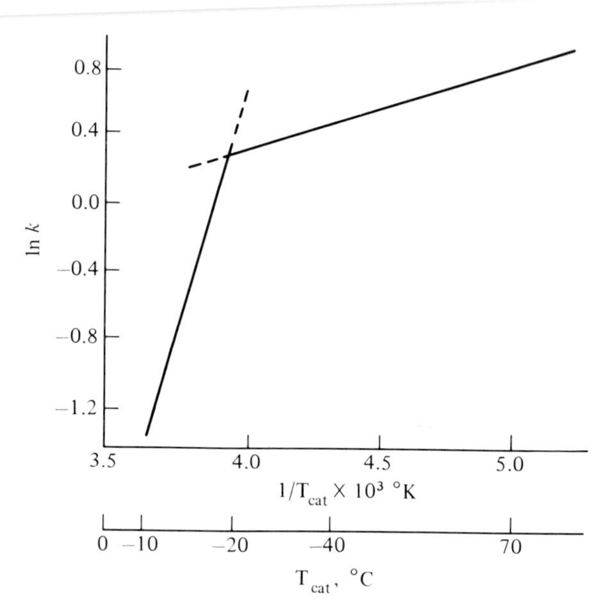

FIGURE 22. Plot of $\ln k = f(1/T_{cat})$ at the optimal value of the molar ratio of the catalyst components ($r = 1$) [101].

3. Correlation between the structure and the technological behaviour of polyisoprene

The molecular and structural properties of a polymer as defined by: molecular weight, molecular-weight distribution, gel content and structure, degree of branching, etc., is of major importance owing to the relationship between the above characteristics and the technological behaviour of the polymer [105].

The technological behaviour of rubbers usually involves all those properties which determine the behaviour of compounds in the various stages of technological processing. The evaluation of all such properties is usually made by using a number of elasto-plastic indicators such as Mooney plasticity and Defo hardness and plasticity, which characterize in fact the effective polymer viscosity under various deformation conditions, chosen as close as possible to those produced in processing operations [14].

The above indicators applied to various known elastomers proved that in certain cases they insufficiently explain the technological behaviour of the given elastomer and usually indicate only the qualitative homogeneity of various batches. As is known, various samples of the same type of elastomer show sometimes a very different technological behaviour although they have the same elasto-plastic indicators [106, 107].

Such peculiarities of elastomers determine their classification into two groups from the point of view of their technological behaviour [108]:

— *non-plasticizable*, i.e. elastomers whose macrostructure is not significantly changed during processing;
— *plasticizable*, i.e. elastomers which change their macrostructure under shear stress. Synthetic polyisoprene belongs to this group. It is characterized by the existence of a supermolecular structure, the gel, which is deeply modified during processing; hence, its technological behaviour is less determined by the original molecular parameters of the polymer than by those following its processing. In such conditions, evaluation of technological properties only by the original elasto-plastic indicators is insufficient and the characteristics of supermolecular structures should therefore also be taken into account. Therefore, the gel poses interesting problems both practical and theoretical, related to the mechanism of its formation during the polymerization process and to its influence on technological behaviour as well [109].

3.1. Mechanism of gel formation in *cis*-1,4-polyisoprene

Gel formation in rubber was first reported in experiments on the emulsion polymerization of butadiene–styrene copolymers. Further studies showed that in such rubbers, the gel involves polymers with a three-dimensional structure formed by crosslinking of the macromolecules within the latex particles. The assumption that the formation of crosslinked polymer would no longer take place in solution polymerization has not been confirmed. *Cis*-1,4-polyisoprene synthesized by polymerization in the presence of certain complex catalyst systems may contain up to 50% of insoluble fractions in static conditions. Efforts were made to explain the structure of such insoluble fractions which, by analogy with the structure of similar fractions in butadiene–styrene rubbers, were initially assumed to be formed of strongly crosslinked macromolecules, called gel [110].

Later research proved the incorrectness of such hypotheses. For instance, the swelling indices can take values up to 50 which are characteristic of loose structures. On the other hand, one has noticed that by vigorously stirring polyisoprene in benzene and filtrating the resulting solutions through filters with pore diameters of 4–5 μm [111], solubilities of almost 100% are obtained. As polymers with such a swelling index cannot pass through the pores of such a filter — given their size — the only explanation could be that during stirring the network is destroyed, which is inconsistent with the three-dimensional structure attributed to the gel. Therefore, it was assumed that the insoluble fraction must be aggregates of single macromolecules attached with one end to a nucleus, a catalyst residue for instance (Figure 23) [84].

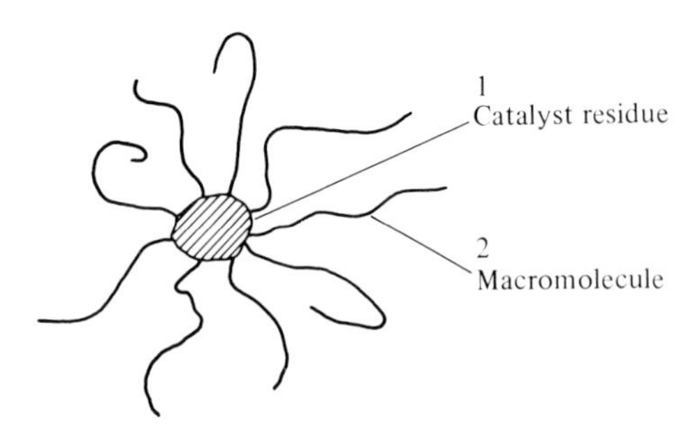

FIGURE 23. Model of a gel structural unit in *cis*-1,4-polyisoprene [84].

Depending on the distance between the nuclei and on the number of molecules involved in an aggregate, the latter can interlace, resulting in a network, at whose junction points either nuclei or physical macromolecular entanglements of various units can be found.

The disadvantage of this gel model arises from the assumption that the star structure is stabilized by the presence of catalyst residues, this is not in agreement with the titanium content of the gel fraction, which is comparable with the content to the soluble fraction.

Therefore, one looks for another explanation still based on the gel star structure, for the stabilizing factor of intermolecular links. Starting from the fact that catalyst systems of $TiCl_4 + AlR_3$ type are active in the cleavage of double bonds via a cationic mechanism especially in the presence of certain impurities with cationic catalytic properties, it was assumed that, depending on the concentration of polymer in the system, two classes of reactions can take place [82, 112, 113]:

— at low polymer concentrations, at a high solvent/polymer ratio, an intramolecular reaction occurs, namely macromolecular cyclization;
— at high polymer concentrations, two competitive reactions occur, i.e. intramolecular cyclization and the linking of different macromolecules on the catalyst surface. In this case, random branched star structures are formed. Their occurrence after deactivation and washing is no longer dependent on the presence of a catalyst residue as the nucleus (Figure 24) [82].

The above-described structure of the gel units can either remain as such or linked to one another to form denser or looser aggregates, depending on the ratio between the average length of the macromolecule and the average distance between the catalyst particles. The possibility of link formation between macromolecules grown on different catalyst particles and linked by physical forces is also to be taken into account.

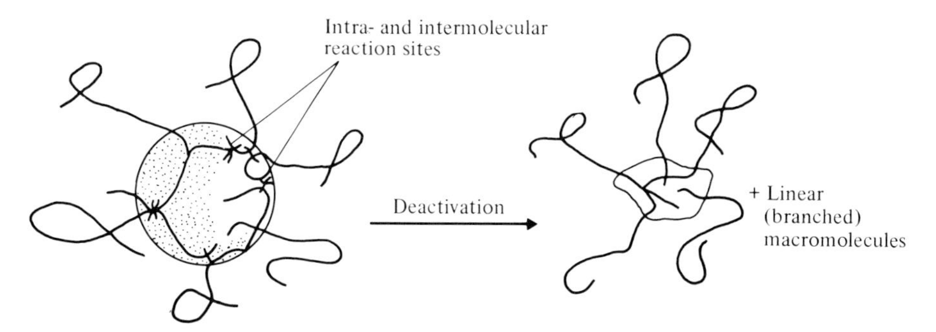

FIGURE 24. Diagram of the macromolecular linking on the catalyst surface and the formation of a gel [82].

The content of polymeric catalyst particles in the polymerization mixture was determined with the ultracentrifuge. It was noticed that towards the end of the polymerization reaction, 50 to 60 % of the macromolecules resulted as aggregates, with the remainder mainly as individual macromolecules. Therefore, the polyisoprene–isopentane system cannot be considered as a true molecular solution. The occurrence of aggregates as three-dimensional networks also causes the anomalous behaviour and the elastic properties of the system (the Weissenberg effect) [81].

It was found that the aggregate content decreases after deactivation, which is also observed on a macroscopic scale by the sharp decrease of the viscosity of the reaction mixture upon the addition of the deactivating agent (see also p. 177). The more dilute the polymerization mixture, the more marked the decrease in viscosity.

The above-mentioned experimental findings are to be explained by the former assumptions on the mechanism of gel formation. Thus, at low polymer concentrations, the number of macromolecules growing on a catalyst particle is usually low and there is a very limited probability of link formation between macromolecules on the same catalyst particle or between one macromolecule grown on one catalyst particle with the growing end of another macromolecule sited on another catalyst particle. By deactivating such systems, almost straight macromolecules and small amounts of aggregates are formed.

At a high polymer concentration, apart from an insignificant decrease in the number of aggregates, it is possible that when aggregate densities are high, the deactivating agent cannot penetrate into the nuclei which may include the catalyst residues. This explains why in the separation of various gel fractions by the ultracentrifuge, the ash content (i.e. the titanium content) of the fractions decreases, as the speed is increased. At high speed, loose gel is separated; from it the catalyst residues are more easily removed by deactivation and washing [109].

Characterization of the aggregate from the point of view of the number of component macromolecules is made by the swelling index, σ (gravimetrically determined in toluene, heptane, hexane or other solvents) [80]:

$$\sigma = \frac{P}{P_0} \quad (12)$$

where P is the weight of the swollen gel, and P_0 is the weight of the swollen gel after drying.

It is considered that at $\sigma \geqslant 20$, the gel is loose and at $\sigma < 15$–20, it is regarded as dense.

Such aggregates, containing a very large number of macromolecules, are difficult if at all, to solubilize and form the insoluble fraction of polyisoprene. They can further become supermolecular structures which in rubber appear as heterogeneous, hard particles [112–114].

Spectroscopic analyses of the structure of various polyisoprene fractions proved that the soluble part (sol fraction and loose aggregates) has the same high content of *cis*-1,4 units, while the dense gel and the solid inclusions contain only 65 and 45 % respectively, of *cis*-1,4 units [115]. The fact that no solid inclusions are formed in the synthesis of polyisoprene without gel is an evidence for the same formation mechanism of these two supermolecular structures [116].

3.2. Influence of gel on the properties of *cis*-1,4-polyisoprene

The complexity of estimating the technological behaviour by the usual elasto-plastic characteristics made us extend our research in order to find the particular factors which affect processing [117–124].

By plotting Mooney viscosity and polyisoprene plasticity *vs.* intrinsic viscosity for both the crude polymer and the soluble fractions (Figure 25) we notice that in the case of crude rubber, no correlation was possible, while for the soluble fractions, the following equations could be derived [120]:

$$\text{ML-2}(100°\text{C}) = 6.14[\eta]^{1.5} \quad (13)$$

$$1/P = 0.83[\eta]^{0.65} \quad (14)$$

where ML is Mooney plasticity in Mooney units,
P is the plasticity (according to GOST 14925-69), and
$[\eta]$ is the intrinsic viscosity in benzene at 20°C, dl/g.

The technological behaviour characteristics of the gel fraction depend on the swelling index [120] (Figure 26).

The technological behaviour of crude rubber depends on several variables, namely the molecular weight of the soluble fraction, the gel content and the swelling index [119] (Figures 27 and 28).

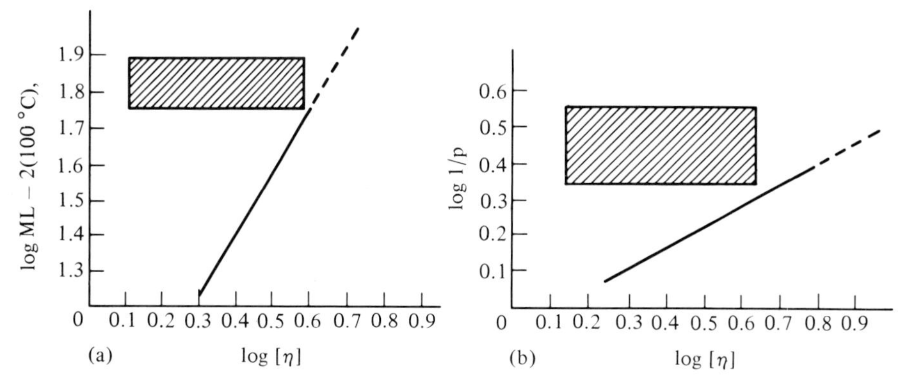

FIGURE 25. Dependence of the Mooney viscosity (a) and plasticity (b) of soluble fractions of *cis*-1,4-polyisoprene on the intrinsic viscosity. The corresponding values of the crude polyisoprene are within the hatched area [120].

The analysis of the above variables led to the following results:

— Mooney viscosity increases, while plasticity decreases with an increase in the gel content of polyisoprene containing a loose gel ($\sigma > 20$). Such correlations become more evident as the molecular weight of the soluble fraction is lower (Figure 27);

— no correlations were observed between the elasto-plastic characteristics and the gel content in polyisoprene containing a dense gel ($\sigma < 20$); Mooney viscosity is higher, while plasticity (GOST) is reduced (Figure 28).

The above investigations could explain the different processing behaviour of polyisoprene batches with similar elasto-plastic indicators. Thus, a crude polyisoprene containing a dense gel and a soluble fraction of relatively low molecular weight may have apparent appropriate values for Mooney viscosity

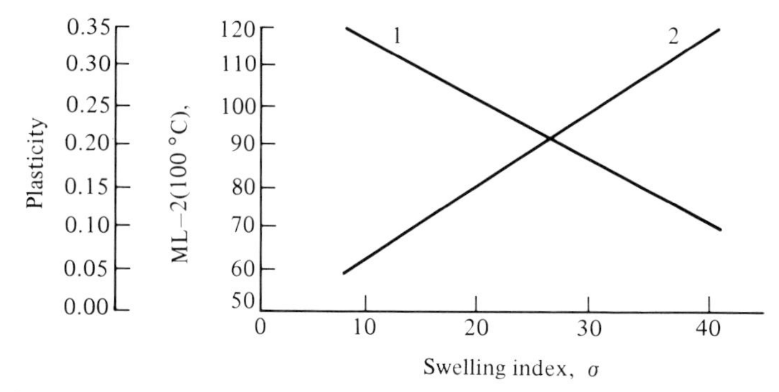

FIGURE 26. Dependence of the Mooney viscosity (1) and plasticity (2) of the gel fraction on the swelling index [120].

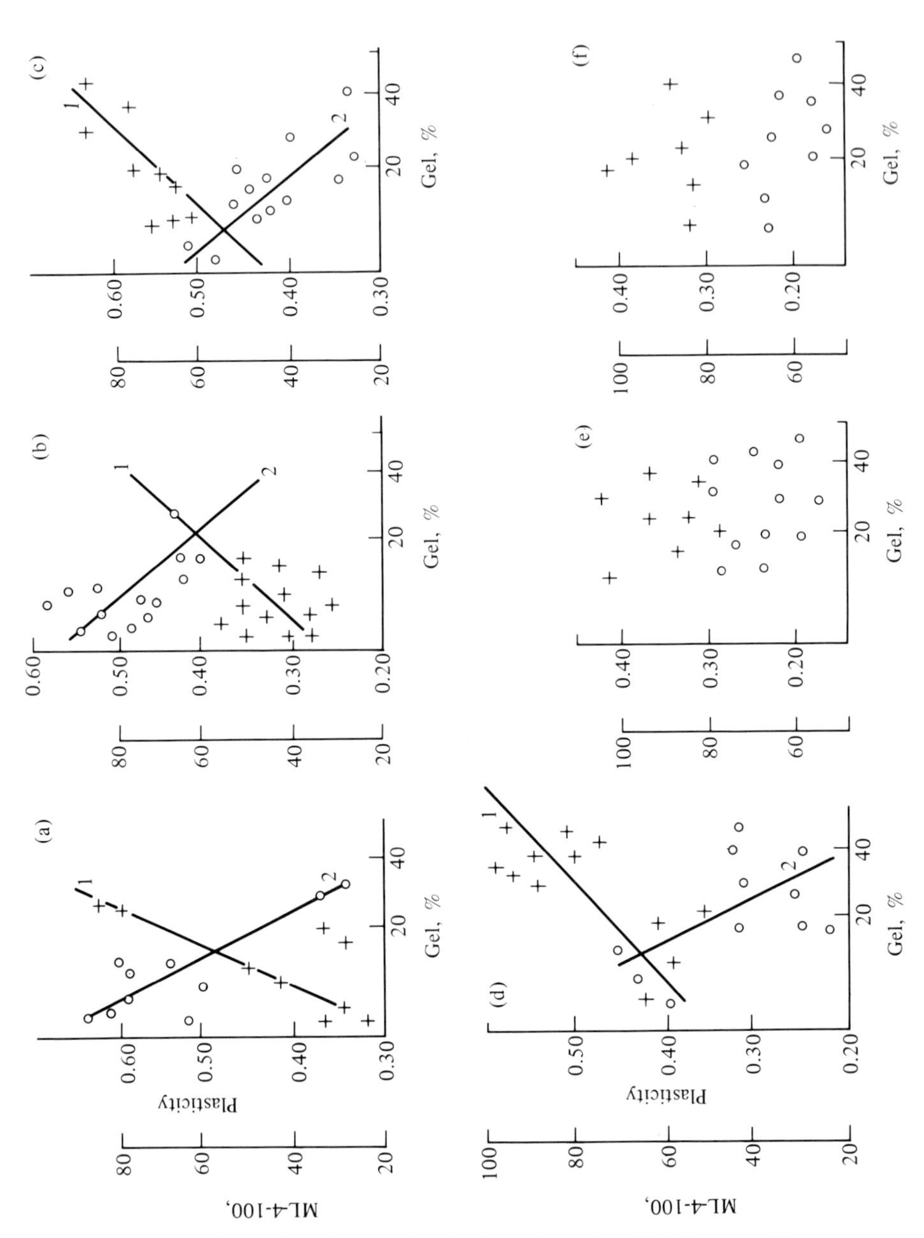

FIGURE 27. Dependence of the Mooney viscosity (+) and plasticity (○) of crude polyisoprene on the molecular weight of the soluble fraction and on the loose gel content. Gel swelling index $\sigma > 20$. Intrinsic viscosity of the soluble fraction in benzene at 20°C, $[\eta]$, dl/g: a = 1.70–2.00; b = 2.01–2.30; c = 2.31–2.60; d = 2.61–3.00; e = 3.01–3.50; f = 3.51–4.00 [119].

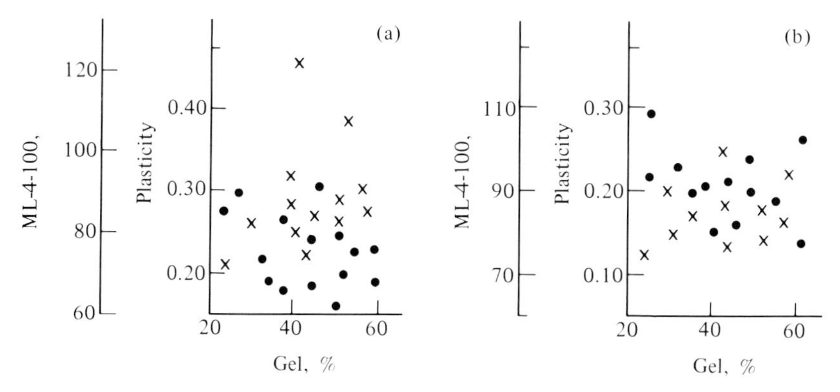

FIGURE 28. Dependence of the Mooney viscosity (×) and plasticity (○) of crude polyisoprene on the molecular weight of the soluble fraction and on the dense gel content. Gel swelling index $\sigma < 20$. Intrinsic viscosity of soluble fraction in benzene at 20°C, $[\eta]$, dl/g: a = 2.61–3.00; b = 3.01–3.50 [119].

and plasticity. In practice, however, it was observed that during mill processing the dense gel is not plasticized and remains as heterogeneous particles, which results in wrinkle formation or even tearing on the mill.

The above properties are not sufficient to characterize the processability of polyisoprene; therefore, other physico-chemical characteristics have to be used, conjointly, such as the intrinsic viscosity of the soluble fraction, the gel content and its swelling index.

Some authors [120] thus recommend an optimum macrostructure for polyisoprene rubber from the point of view of its technological properties, namely:

—a content of soluble fraction of 75–85%, with an intrinsic viscosity in benzene at 20°C, $[\eta] = 3.3\text{–}3.8$ dl/g;
—a content of gel fraction of 15–25%, with a swelling index, $\sigma \geqslant 25$.

We have also noticed the existence of an optimum macrostructure and of some physico-mechanical characteristics [109]. As shown in Figure 29, the tensile strength is a maximum at a gel content of about 20–30% for a polyisoprene with a swelling index $\sigma > 25$ (curve 1), while at a swelling index $\sigma < 25$, the tensile strength continuously decreases with an increased gel content (curve 2).

A process which would yield a homogeneous product from the point of view of the above-mentioned indicators is virtually impracticable as it would involve the concomitant adjustment of three variables which are difficult to measure continuously.

Bearing in mind that the technological properties of gel-free polyisoprene rubber can only be adjusted by the value of the molecular weight, we studied

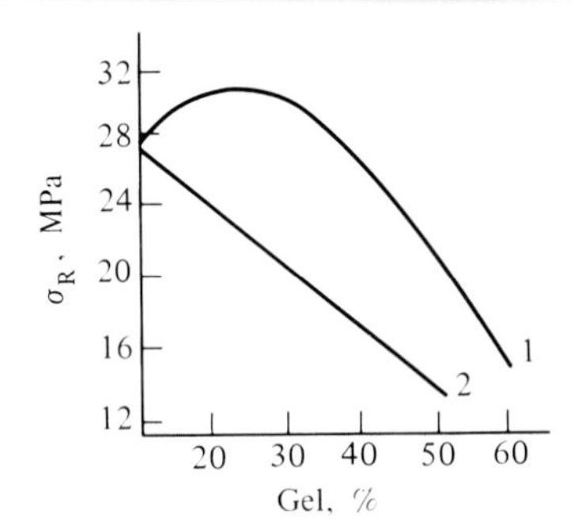

FIGURE 29. Influence of gel content on the tensile strength of *cis*-1,4-polyisoprene. 1. Polyisoprene with a swelling index >25; 2. polyisoprene with a swelling index <25.

the conditions for the synthesis of a rubber with no gel fraction, the molecular weight of which could be adjusted with an agent for controlling molecular weight. Taking into account that cationic reactions produced on the catalyst surface are responsible for gel formation, it was suggested that they should be avoided by a stepwise addition of the monomer to the reaction zone containing the catalyst, the result being a gel-free polymer [116, 119].

The use of hydrogen was suggested for regulating the molecular weight, providing an efficient means for the control of the final product [125, 126]. In such conditions, the variables are controlled by a computer fed with a mathematical model of the process, where signals from the flow rate, temperature and reaction viscosity transducers (proportional to the molecular weight and concentration of *cis*-1,4-polyisoprene) are processed [127–129].

Chapter III

Deactivation of the catalyst complex

THE polymer solution resulting from the stereospecific polymerization reaction of isoprene still contains the catalyst complex used in an active form. Experimental data indicate that the best physico-mechanical properties of the polymer are obtained at a certain optimum conversion which is lower than the maximum level attainable [50]. As previously indicated, the optimum conversion takes higher values in the presence of modified catalyst complexes. However, for economic reasons, the polymerization reaction must be carried out at a maximum conversion to simplify the recovery of the unreacted monomer. When optimum or maximum conversion is reached, the polymerization reaction is terminated by destroying the catalytic activity of the system used. The deactivation step is necessary since the active catalyst form, β-$TiCl_3$, reacts with oxygen at room temperature, resulting in the formation of peroxides like $Cl_3Ti{-}O{-}O{-}TiCl_3$. Such peroxides are easily decomposed to free radicals, such as $Cl_3TiO\cdot$, which further initiate polymer degradation [130, 131]. Intensive studies have emphasized the major role of transition metals in the polymer degradation process [132–134].

Numerous methods of deactivating the catalyst complex and removing or deactivating the resulting catalyst residues have been suggested so as to obtain a polymer with the lowest possible ash content, thus maintaining its physico-chemical and physico-mechanical properties during conditioning, storage and processing.

The methods of deactivation of the catalyst systems used in the synthesis of polyisoprene are classified into the following three groups, using as a criterion the active or inactive form of the transition metal after deactivation:

(1) deactivation of the catalyst complex with compounds which do not inactivate the transition metal, followed by washing out the catalyst residues. In such cases, the removal of catalyst residues from the polymer is absolutely necessary, as their presence in it causes discolouration and is detrimental to the physico-chemical and physico-mechanical properties, because of their catalytic effect in the complex degradation process;

(2) deactivation of the catalyst complex with compounds which inactivate the transition metal;
(3) deactivation of the catalyst complex via a combined method.

1. Deactivation of the catalyst complex without inactivation of the transition metal

Such catalyst complex deactivation is achieved with compounds containing functional groups, i.e. alcohols and carbonyl compounds.

(a) *Deactivation of the catalyst complex with alcohols*

C_1–C_5 alcohols are usually preferred. They are used in amounts of 1 to 10 % by volume [135]. In the deactivation of a Ziegler–Natta-type catalyst complex with alcohol in an inert medium, the reaction occurs instantaneously and is associated with a sudden change in colour [136]:

$$TiCl_3{\cdot}AlR_2Cl + 8R'OH \rightarrow Al(R'O)_2Cl + 2RH + TiCl_3{\cdot}6R'OH \quad (1)$$

The above equation accounts for the stoichiometric amount of alcohol needed in the deactivation of the catalyst system. Titanium hexa-alkoxides are soluble in excess alcohol, apart from the derivative obtained with isopropyl alcohol [136].

Titanium hexa-alkoxides can more rapidly react with oxygen than β-$TiCl_3$, even at temperatures of 18–25°C, yielding peroxides with the following formula: $Cl_3Ti{-}O{-}O{-}TiCl_3$. Their major role in polymer degradation was previously mentioned. Therefore, the removal of catalyst residues from the polymer by washing it with excess alcohol or with water is imperative [131, 136].

(b) *Deactivation of the catalyst complex with carbonyl compounds*

Deactivation with carbonyl compounds is quite similar to deactivation with alcohols. The most widely used carbonyl compound is acetone [137]. The same stoichiometric amount of deactivating agent is needed as in deactivation with alcohols, the reaction being similar.

2. Deactivation of the catalyst complex with inactivation of the transition metal

This catalyst complex deactivation method can be achieved with basic nitrogen compounds or with chelating agents.

(*a*) *Deactivation of the catalyst complex with basic nitrogen compounds*

Deactivation of the catalyst complex with basic nitrogen compounds, mainly reported in the patent literature, involves the conversion of the catalyst into inactive, insoluble or soluble, hardly oxidizable and hydrolizable compounds which no longer catalyse polymer degradation during the further processing steps [138]. Apart from gaseous or liquid ammonia, aniline and pyridine are also used. All the above nitrogen compounds decompose the catalyst via the formation of a complex; extraction of the catalyst residues by washing is no longer necessary in this case [136, 139].

(*b*) *Deactivation of the catalyst complex with chelating agents*

As is well-known, chelates are a class of metallic compounds where a metal atom or ion is attached by several chemical bonds to the same molecule (usually an organic compound), called a chelating agent [140]. The deactivation of catalyst complexes used in the stereospecific polymerization of isoprene is based on the ability of a transition metal, namely titanium in this particularly case, to form chelates in the presence of chelating agents. Characteristic of a chelating agent is the presence of one or more atoms, called donors, with unshared electron pairs available for the formation of a coordinate bond. The specific effect of chelate formation is the chemical inactivation of the metal. This effect is proportional to the number of bonds formed by the chelating agent. The highest degree of inactivation is achieved by complexing the metal in its highest coordination state [141]. In such cases, the catalyst not only becomes inactive in oxidation, but it even acts as an inhibitor of the oxidation reaction [142].

Complex formation of polyvalent metals in their lower coordination state leads to the formation of compounds able to decompose hydroperoxides; as a result the chelates have a pro-oxidant effect [143].

Formation of the transition metal complex is sometimes made after its conversion to its highest valency state. Compounds such as nitrobenzene, azobenzene or benzoquinone are used to oxidize Ti^{3+} to Ti^{4+} according to the following equations [131]:

$$6TiCl_3 + C_6H_5NO_2 + 6HCl \rightarrow 6TiCl_4 + C_6H_5NH_2 + 2H_2O \quad (2)$$

$$4TiCl_3 + C_6H_5N{=}N{-}C_6H_5 + 4HCl \rightarrow 4TiCl_4 + 2C_6H_5NH_2 \quad (3)$$

$$2TiCl_3 + O{=}C_6H_4{=}O + 2HCl \rightarrow 2TiCl_4 + HO{-}C_6H_4{-}OH \quad (4)$$

The stability of the resulting chelates largely depends on the efficiency of the various bonds. A major role is also played by the arrangement of the ligands around the metal ion.

Phosphoric acid, an inorganic chelating agent, used in the deactivation of Ziegler–Natta-type catalyst systems leads to the formation of four-membered cyclic chelates where titanium is complexed in its highest coordination state and is attached to three phosphate groups [126].

Among the organic compounds which contain functional groups and atoms with unshared electron pairs available for the formation of coordinate bonds, EDTA (disodium salt of ethylenediaminetetraacetic acid) is the chelating agent most frequently used.

Certain polymers are known to behave as chelating agents, such as the homopolymer of diallyl-(3-formyl-4-hydroxybenzyl)methylammonium chloride or its copolymer with SO_2 [144].

(c) *Deactivation of the catalyst complex by combined methods*

When the chelating agent is expensive, a combination of the two above-mentioned methods is used to deactivate the catalyst complex. Such a combined method is reported by Rosik and Svoboda [136] who decomposed the catalyst system with a lower alcohol, removed some of the catalyst residue from the polymer and that part of the catalyst residue which could not be removed was converted into a chelate with phosphoric acid. The method is less satisfactory than the directly chelating one, as the catalyst residues are still capable of the catalytic oxidation of the polymer [136].

3. Deactivation of the catalyst complex in the absence of polyisoprene

The most important factors which have a decisive effect on the course of deactivation of the catalyst complex are related to the nature and amount of the deactivating agent, the viscosity of the polymer solution, the stirring, and the time and temperature at which deactivation is undertaken.

We first carried out the deactivation and then removed the catalyst residue by washing in order to achieve a low ash and transition metal content. This procedure was satisfactory for polymerization processes involving a relatively high catalyst consumption. Studies on deactivation with chelating agents were later initiated together with investigations on the more active catalyst systems which permitted lower unit consumptions of catalyst.

As reported in the literature, deactivating compounds lead to acidic or basic reactions, depending on their nature. Basic type deactivating agents form basic salts of aluminium and titanium or even hydrated aluminium oxide. Such substances are insoluble and precipitate on the surface of the polymer particles resulting in a rubber with a high ash content. Acidic or neutral-type deactivating agents, in the presence of strong or weak acids, form soluble aluminium and titanium salts (e.g. chlorides) easily removed from the polymer solution by washing with water. Therefore acidic-type deactivation is preferred.

The deactivating agents which instantaneously form easily removable soluble salts are the most convenient for the $TiCl_4 + AlR_3$ catalyst system. The deactivating agents must be soluble both in the reaction medium and in water. In the presence of the two phases (reaction medium and water), its partition coefficient, as well as that of the products formed by deactivation of the catalyst complex, should be higher in the aqueous phase.

Deactivation and washing is a very complex problem owing to the high viscosity of the polymer solution, the presence of two phases and the influence of atmospheric oxygen. An intimate contact between the deactivating agent and the polymer solution must be reached, in the first place.

A study of the behaviour of the catalyst complex during deactivation was initially made in the absence of polymer.

The behaviour of the catalyst complex in the presence of various deactivating agents was investigated experimentally and certain conclusions were drawn as to the course of the reaction. The rate of disappearance of the brown colour indicated the rate of decomposition of the catalyst complex. In the next step, still in the absence of the polymer and in an inert medium, the washing was with distilled and/or acidified water, further noting the colour changes of the medium, the solubility of the decomposition residues in the aqueous and hydrocarbon phases, the tendency for the formation of an emulsion, and the time needed for phase separation.

The presence of titanium in the two phases was tested by means of a colour reaction with a reagent containing H_2O_2 in H_2SO_4 which developed a yellow colour easily observed in the presence of traces of titanium.

As deactivating agents alcohols, ketones, acids, and amines, either alone or in mixtures, were used.

When alcohols were used, the deactivation time increased with the number of carbon atoms in the molecule. The lower the alcohol, the more easily was the emulsion formed by the two phases broken and separated. Regardless of the alcohol used, titanium and aluminium were almost quantitatively extracted from the organic phase by a small number of washings with water. In an acidic medium, catalyst deactivation with alcohols occurred in a similar way although the phase separation was faster.

When alcohols and organic acids were used, the phase separation was fast. When higher or polybasic acids were used, a precipitate of TiO_2 was however formed.

Deactivation of the catalyst complex with acetone and hydrogen chloride occurred almost instantaneously and phase separation during washing was very simple; aluminium and titanium were quantitatively extracted.

In the series of aliphatic amines tested, ethylenediamine deactivated the catalyst complex at a higher rate, but with diethanolamine, the removal of catalyst residues was almost quantitatively achieved. Pyridine did not give good results, as it formed a persistent emulsion.

Other compounds such as hydrogen peroxide and hydrogen chloride or the disodium salt of tartaric acid were also effective in the deactivation of the catalyst complex, but owing to certain secondary effects, e.g. stable emulsions in the case of hydrogen peroxide and precipitate formation with the sodium salt of tartaric acid, their use is not recommended.

4. Deactivation of the catalyst complex in the presence of polyisoprene

Based on the observations during this study on the rate at which the deactivating agent decomposes the catalyst, on the solubility of the deactivating agent and of the products formed in the decomposition of the catalyst, on the behaviour during the washing step and on the separation of the two phases, further studies were made on the deactivation of the catalyst complex in the presence of polyisoprene.

The deactivation process is much more complex in this case and its investigation involves even more difficulties. The high viscosity of the polymer solution makes an intimate contact between the deactivating solution and the washing agent impossible. A longer contact time, higher amounts of deactivating agents and longer periods for phase separation were necessary.

Although both the deactivation and washing steps were carried out under an inert-gas, a concomitant polymer stabilization is advisable. The polymer properties during deactivation and washing are preserved. However, this method has the disadvantage that part of the stabilizer may be lost during washing. A method with no such disadvantage and which nevertheless ensures polymer stabilization, consists in the addition of a part of the stabilizer with the deactivating agent and the rest after the polymer solution has been washed.

In order to estimate the efficiency of the deactivation process, the variation of the polymer ash content and the presence of titanium in the washing water by qualitative determinations with H_2O_2 and H_2SO_4 were observed.

Among the numerous deactivating agents, the following were chosen: isopropyl alcohol, methyl alcohol, ethyl alcohol, acetone, triethanolamine and hydrogen peroxide; these were tested for their ability to deactivate the catalyst complex in the presence of a polymer.

At first, only water acidified with hydrochloric acid was used for washing, which ensured an acidic medium and facilitated solubilization of products formed in deactivation. Distilled water was later used until the original pH of the washing water was reached.

The efficiency of the various deactivating agents was determined in virtually identical experimental conditions, i.e.

—amount of deactivating agent: 30 moles/mole of catalyst;
—reaction temperature: 20°C;
—contact time: 30 minutes;
—viscosity of reaction mixture: about 3000 cP;
—intrinsic viscocity of polymer: $[\eta] \sim 3.0$ dl/g;
—amount of washing water for a single washing: 50% of the reaction mixture by volume;
—stirrer speed: 150 rpm.

The results are given in Table 10. Methyl alcohol, ethyl alcohol and acetone were almost equally efficient as deactivating agents; the content of inorganic compounds in the polymer reached an acceptable concentration after four washings.

With isopropyl alcohol, the phase separation needed a longer time, i.e. more than 1 hour; titanium salts were in fact not extracted in the washing water and the polymer ash content was consequently very high.

In deactivation with triethanolamine, although the polymer ash content was below the standard values, the washing stages needed a long time, the phase separation being very difficult. When hydrogen peroxide was used a stable emulsion resulted, not separated even after 8 hours.

Based on the above results, in order to investigate further the deactivation step, two agents, i.e. methyl alcohol and acetone, were chosen, which were more suitable from both the deactivating efficiency and the technological point of view. We attempted to find the optimum amount of deactivating agent and the optimum contact time between the deactivating agent and the polymer solution to achieve completion of the reaction. We also assessed the influence of the viscosity of the reaction medium, the number of washings required and the quantity of washing agent needed to achieve an adequate removal of the catalyst residues.

According to the reversible reaction between the catalyst system components used in polymerization and an alcohol or a ketone, 8 moles of deactivating agent per mole of catalyst are needed. However, as reported in the literature [136] and as a result of our previous investigations on the deactivation of the catalyst complex in the absence of a polymer, no reproducible data were obtained under the above conditions which could be explained by a different reaction rate of the catalyst components with alcohol or ketone. The aluminium derivatives in solution react first as they are more reactive. $TiCl_3$ is able to react only if the amount of deactivating agent is increased. If the deactivating agent is in a large excess, a solvation layer is also formed which increases the solubility in water of the resulting products.

Table 10

Variation of ash content of polyisoprene as a function of the nature of the deactivating agent

No.	Deactivating agent	Number of washings	Ash content %	Remarks
1	Isopropyl alcohol	6	0.8	Difficult separation
2	Isopropyl alcohol	6	0.76	idem
3	Ethyl alcohol	6	0.4	Good separation
4	Ethyl alcohol	6	0.4	idem
5	Methyl alcohol	4	0.25	Very good separation
6	Methyl alcohol	6	0.3	idem
7	Acetone	4	0.3	idem
8	Acetone	6	0.3	idem
9	Triethanolamine	6	0.4	Difficult separation
10	Hydrogen peroxide	6	—	Stable emulsion

Efficient reproducible results were obtained by using 30 moles of acetone or alcohol per mole of catalyst.

The presence of the polymer in the system greatly increases the viscosity, and the catalyst particles imbedded in the polymer react more slowly. Therefore, larger amounts of deactivating agent are expected to be needed in the presence of polymer.

In experiments carried out to establish the optimum amount of alcohol or acetone, the viscosity of the polymer solution, as well as the speed of the stirrer, the temperature and the time of the deactivation reaction were kept constant.

The results are shown in Table 11.

Table 11
Effect of amount of deactivating agent on the ash content of the polymer after washing

No.	Deactivating agent	mole/mole of catalyst	Ash content of the final product (%)
1	Methyl alcohol	10	0.45
2	Methyl alcohol	10[a]	0.4
3	Methyl alcohol	30	0.4
4	Methyl alcohol	30[a]	0.35
5	Methyl alcohol	50	0.20
6	Methyl alcohol	50[a]	0.18
7	Methyl alcohol	100	0.25
8	Methyl alcohol	100[a]	0.20
9	Methyl alcohol	200	0.20
10	Methyl alcohol	200[a]	0.19
11	Acetone	10	0.55
12	Acetone	10[a]	0.55
13	Acetone	30	0.4
14	Acetone	30[a]	0.35
15	Acetone	50	0.24
16	Acetone	50[a]	0.20
17	Acetone	100	0.25
18	Acetone	100[a]	0.25
19	Acetone	200	0.3
20	Acetone	200[a]	0.25

[a] First two washings with acidified water.

No significant differences between the two deactivating agents were found. The ash content of the polymer decreases with an increased amount of deactivating agent from 10 to 50 moles per mole of catalyst, remaining afterwards at an almost constant level.

Runs no. 11-4 and 11-14, respectively, needed a great number of washings, the extraction of catalyst residues by the washing water being more difficult even when the first two wash waters were acidified. By using 10 and 30 moles of deactivating agent per mole of catalyst, the amount of ash formed in the polymer was not reduced to below the permitted limits. One can conclude that for an efficient deactivation, a ratio of 50 moles of deactivating agent per mole of catalyst is needed.

The contact time between the deactivating agent and the polymer solution, an important factor in the deactivation of the catalyst system, is largely determined by the achievement of a more efficient contact surface between the deactivating agent and the components of the catalyst system in the viscous solution of the polymer. We worked at a stirring speed of 900 rpm. The viscosity of the polymer solution was kept constant at about 3500 cP and the amount of deactivating agent was 50 moles per mole of catalyst in all runs. Washing was carried out with a water volume of 50% of the volume of the polymer solution and the first two wash waters were acidified. The contact time between the deactivating agent and the polymer solution was varied, the colour change of the solution was observed and the ash content of the polymer was determined.

The results are given in Table 12. It can be seen that 10 minutes are not sufficient for a complete deactivation of the catalyst complex. A contact time of 30 minutes is enough to achieve a complete deactivation but a longer contact time, e.g. 90 minutes, does not improve the process.

Table 12

Effect of the deactivating agent type and of the deactivation time on the ash content of the polymer

(deactivating agent concentration 50 moles/mole catalyst)

No.	Deactivating agent	Deactivation time, min	Ash content, %
1	Methanol	10	0.6
2	Methanol	30	0.2
3	Methanol	60	0.2
4	Methanol	90	0.18
5	Acetone	10	0.62
6	Acetone	30	0.2
7	Acetone	60	0.2
8	Acetone	90	0.2

Since the viscosity of the polymer solution changes, we studied several polymer solutions with various viscosities, treating them identically in order to deactivate them and remove the soluble catalyst residues by washing. The results of such experiments are listed in Table 13.

One observes that in the viscosity range studied from 1520 to 5760 cP the deactivation of the catalyst complex is not affected by the type of deactivating agent used.

The ash content of the final polymer has a negative effect on its properties; hence, the removal of soluble catalyst residues by washing the polymer

Table 13
Effect of viscosity of polymer solution on the deactivation process

No.	Polymer solution viscosity (cP)	Deactivating agent	Ash content (%)
1	5760	Acetone	0.15
2	2870	Acetone	0.13
3	1520	Acetone	0.20
4	5760	Methanol	0.13
5	2870	Methanol	0.13
6	1520	Methanol	0.18

solution is highly important. We tried to establish the number of washings needed for a complete removal of such residues.

The working conditions were the same in all runs, i.e. we used methyl alcohol as deactivating agent (50 moles/mole of catalyst), a deactivating time of 30 minutes, distilled water as the washing agent, and a washing time of 20 minutes. The results are shown in Table 14.

The presence of titanium was determined in the wash water by using the reaction with H_2O_2 and H_2SO_4 as a qualitative test; after washings nos. 4 and 5, the water contained no more titanium. Table 14 indicates that after the fourth washing, the ash content remained constant. If a volume of water of 50 % of the volume of the polymer solution was used in each washing, the same efficiency was reached as for a water volume of 80 %. The conclusion is drawn that washings with lower quantities of water are more economic, the number of washings being maintained but the total quantity of wash water is much reduced.

In order to establish the influence of catalyst residues on the stability of *cis*-1,4-polyisoprene, the oxygen-absorption isotherms were determined on two polyisoprene samples, A and B, isolated from the same original solution, which had been treated with methanol to deactivate the catalyst complex. Sample A was obtained by separating polyisoprene without removing the catalyst residues, while sample B was isolated from the same solution after the removal of those residues. The results are shown in Table 15 and in Figure 30.

The following conclusions have been drawn by analysing the oxygen-absorption isotherms at various temperatures, as shown in Figure 30.

— the curve shape is characteristic of an oxidation process proceeding via an autocatalytic free radical mechanism;
— the overall amount of absorbed oxygen does not depend on temperature;
— the oxidation induction period, τ, is dependent on temperature, decreasing with an increase in temperature.

Table 14

Effect of number of washings and of the amount of water on the ash content of the final polymer

No.	Initial ash (%)	Washing 1		Washing 2		Washing 3		Washing 4		Washing 5	
		V water/ V solution	Ash (%)	V water/ V solution	Ash (%)	V water/ V solution	Ash (%)	V water/ V solution	Ash (%)	V water/ V solution	Ash (%)
1	1.46	0.5	0.90	0.5	0.47	0.5	0.30	0.5	0.19	0.5	0.18
		0.8	0.80	0.8	0.45	0.8	0.30	0.8	0.20	0.8	0.18
2	1.09	0.5	0.68	0.5	0.50	0.5	0.28	0.5	0.16	0.5	1.16
		0.8	0.60	0.8	0.50	0.8	0.25	0.8	0.19	0.8	0.17
3	1.23	0.5	0.70	0.5	0.45	0.5	0.25	0.5	0.20	0.5	0.20
		0.8	0.75	0.8	0.40	0.8	0.25	0.8	0.19	0.8	0.18

Table 15

Oxygen absorption isotherms as a function of temperature and ash content of polyisoprene

			Oxygen-absorption isotherms			
					Maximum oxygen absorption	
No.	Sample	Ash content (%)	Temperature (°C)	Induction period (min)	Time (min)	cm^3 oxygen/g polymer
1	A	1.23	140	10	220	245
			130	26	295	245
			120	31	390	245
2	B	0.18	140	15	315	245
			130	40	375	245
			120	117	460	245

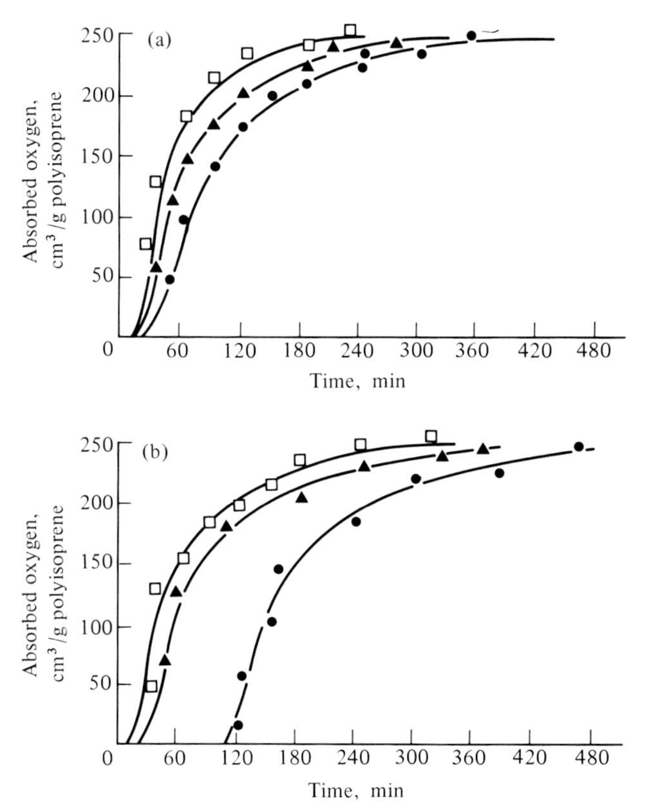

FIGURE 30. Oxygen absorption isotherms of *cis*-1,4-polyisoprene. □ — 140°C; △ — 130°C; ● — 120°C. A — before removal of catalyst residues; B — after removal of catalyst residues.

From a comparison between the oxygen-absorption isotherms of the two *cis*-1,4-polyisoprene samples, we observed that sample A was more rapidly oxidized than sample B, from which the catalyst residues were removed. This fact is outlined by both the induction period and the time needed for the final oxygen-absorption value to be reached. Such experimental data on the influence of catalyst residues present in *cis*-1,4-polyisoprene led to the conclusion that it was necessary to remove the catalyst residues from the polymer below a certain limit by washing with water. The above results also directed research towards finding certain deactivating agents able to form a complex with the catalyst, converting it into a product inactive in oxidation, in order to avoid the washing step.

Out of the deactivating agents which do not inactivate the catalyst residues, both methyl alcohol and acetone proved to be suitable. Methyl alcohol is preferable since, unlike acetone, it can be completely recovered from the washing water and the polymerization solvent. This fact is highly important for the industrial units where the polymerization solvent is recycled; its enrichment in acetone must be avoided, as it deactivates the polymerization catalyst.

Chapter IV

The stabilization of polyisoprene

IT IS well-known that polyisoprene with a high content of *cis*-1,4 addition product is a synthetic elastomer with physico-chemical and physico-mechanical properties closely resembling natural rubber and it can therefore be used as a substitute for natural rubber.

The main difference between synthetic polyisoprene and natural rubber on the one hand, and other similar elastomers on the other, is the sharp decrease in physico-mechanical properties as the result of the reaction of such elastomers with relatively low amounts of oxygen.

Synthetic polyisprene is an elastomer which contains double bonds both in the main chain and in branchings as a result of 1,4-, 1,2- or 3,4-addition of the isoprene, respectively, as shown in the following formulae:

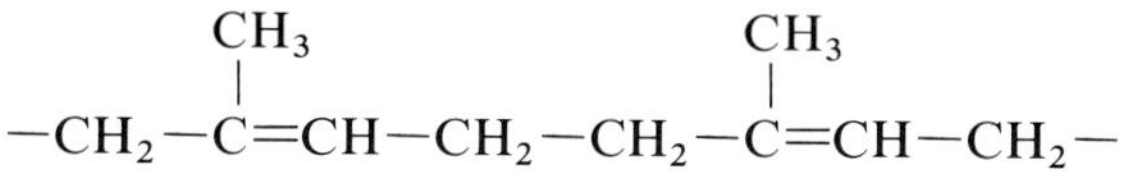

$$-CH_2-\overset{\displaystyle CH_3}{\overset{|}{C}}=CH-CH_2-CH_2-\overset{\displaystyle CH_3}{\overset{|}{C}}=CH-CH_2- \qquad (1)$$

cis- and *trans*-1,4-addition

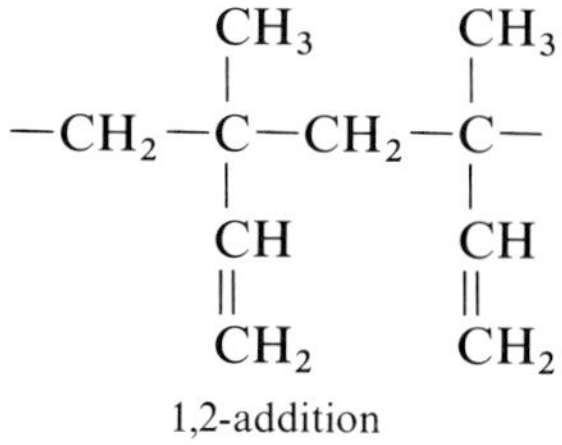

$$\begin{array}{ccccc} & CH_3 & & CH_3 & \\ & | & & | & \\ -CH_2- & C & -CH_2- & C & - \\ & | & & | & \\ & CH & & CH & \\ & \| & & \| & \\ & CH_2 & & CH_2 & \end{array} \qquad (2)$$

1,2-addition

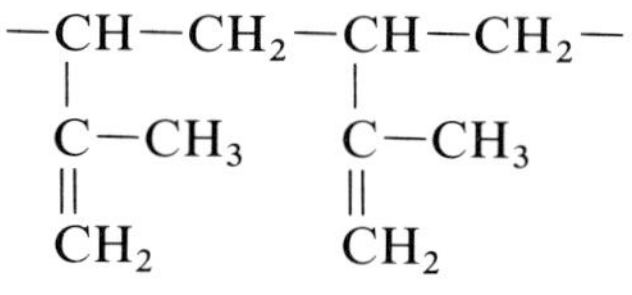

$$\begin{array}{llll} -CH- & CH_2- & CH- & CH_2- \\ \;\;\; | & & \;\; | & \\ \;\;\; C-CH_3 & & \;\; C-CH_3 & \\ \;\;\; \| & & \;\; \| & \\ \;\;\; CH_2 & & \;\; CH_2 & \end{array} \qquad (3)$$

3,4-addition

The olefinic double bonds in the main polymeric chain favour instability of the polymer towards oxygen, ozone, light, etc. This instability is made evident by changes in the physico-mechanical properties of both crude and cured rubber, ultimately affecting the service life of rubber goods.

Unlike natural rubber, which contains a non-rubbery material including compounds with protecting properties, polyisoprene not only does not contain such substances, but has catalyst residues which, depending on the deactivation method, can even catalyse the degradation process. Taking into account the double bond content of the polymer on the one hand and the presence of polyvalent metals as catalyst residues on the other hand, it is obvious that the deactivation step has to be followed by a stabilization step. This consists mainly in the addition of a compound, called antioxidant, which confers stability to the polymer during both the synthesis and the storage stages and later on during processing and use.

The criteria of selecting the best antioxidant (its nature and optimum concentration) obviously depend upon the mechanism of reaction with oxygen.

1. Autoxidation mechanisms

Oxidation of *cis*-1,4-polyisoprene is an autocatalytic process in which hydroperoxides (the main reaction products) are decomposed resulting in the formation of free radicals which initiate a chain reaction. The amount of the absorbed oxygen with time is described by an S-shaped curve, characteristic of an autocatalytic reaction [145, 146].

The following mechanism is accepted for the autoxidation of a hydrocarbon, in the absence of initiators and inhibitors [145, 147]:

initiation:

$$\mathrm{ROOH} \rightarrow \mathrm{RO}\cdot + \mathrm{HO}\cdot \tag{4}$$

$$\mathrm{2ROOH} \xrightarrow{k_i} \mathrm{RO}\cdot + \mathrm{RO_2}\cdot + \mathrm{H_2O} \tag{5}$$

propagation:

$$\mathrm{RO_2}\cdot + \mathrm{RH} \xrightarrow{k_p} \mathrm{ROOH} + \mathrm{R}\cdot \tag{6}$$

$$\mathrm{R}\cdot + \mathrm{O_2} \xrightarrow{\text{fast}} \mathrm{RO_2}\cdot \tag{7}$$

termination:

$$\mathrm{2R}\cdot \rightarrow \mathrm{R{-}R} \tag{8}$$

$$\mathrm{R}\cdot + \mathrm{RO_2}\cdot \rightarrow \mathrm{RO_2R} \tag{9}$$

$$\mathrm{2RO_2}\cdot \xrightarrow{k_t} \text{non-radical products} + \mathrm{O_2} \tag{10}$$

Hydroperoxide decomposition can be initiated by polyvalent metals, light, ozone, etc. Among the above-mentioned factors, the polyvalent metals should primarily be considered since, as previously shown, synthetic polyisoprene includes larger or smaller amounts of such metals. The theoretical background for the participation of metals in the decomposition of hydroperoxides have been established by Uri [148].

When a metal has two valency states with comparable stability (i.e. which can easily undergo oxidation-reduction reactions, either by losing or gaining an electron), the following reactions can take place:

$$ROOH + M^{n+} \rightarrow RO\cdot + M^{(n+1)+} + HO^- \qquad (11)$$

$$ROOH + M^{(n+1)+} \rightarrow RO_2\cdot + M^{n+} + H^+ \qquad (12)$$

$$2ROOH \xrightarrow{M^{n+}/M^{(n+1)+}} RO\cdot + RO_2\cdot + H_2O \qquad (13)$$

From the above reactions, one notices that a relatively small amount of polyvalent metal can transform a considerable quantity of hydroperoxide into free radicals with a fast reaction rate, even at room temperature.

The addition of an antioxidant changes the above kinetic scheme since it participates in the course of the reaction.

The most correct classification of antioxidants is based on their participation in the oxidation reaction under the most severe conditions, i.e. thermal oxidation. According to this criterion, antioxidants are classified into [141]:

(a) preventive antioxidants which inhibit or retard the formation of free radicals in the initiation steps;
(b) chain-terminating antioxidants, which stop the propagation by reaction with the free radicals $R\cdot$ or $RO_2\cdot$.

Some antioxidants may react with oxygen, in this way contributing to the formation of free radicals which are able to initiate the oxidation chain reaction at a later stage [141, 149, 150].

Therefore, the suggested autoxidation mechanism in the presence of antioxidants is as follows [141]:

— initiation
  — peroxide decomposition $\quad nROOH \rightarrow RO\cdot; RO_2\cdot \quad (14)$
  — oxygen attack on the polymer chain $\quad RH + O_2 \rightarrow R\cdot + HO_2\cdot \quad (15)$
  — oxygen attack on the antioxidant (AH) $\quad AH + O_2 \rightarrow A\cdot + HO_2\cdot \quad (16)$

— propagation
  — similar to uninhibited autoxidation $\quad RO_2\cdot + RH \rightarrow ROOH + R\cdot \quad (17)$
  $\quad R\cdot + O_2 \rightarrow RO_2\cdot \quad (18)$

— chain transfer
  — to antioxidant $$RO_2\cdot + AH \rightarrow RO_2H + A\cdot \quad (19)$$
  — to polymer $$A\cdot + RH \xrightarrow{O_2} AO_2H + RO_2\cdot \quad (20)$$

— termination
  — by antioxidant $$RO_2\cdot + A\cdot \rightarrow RO_2A \quad (21)$$
$$2A\cdot \rightarrow A{-}A \quad (22)$$
  — similar to uninhibited autoxidation $$2RO_2\cdot \rightarrow \text{non-radical products} \quad (23)$$
$$RO_2\cdot + R\cdot \rightarrow ROOR \quad (24)$$
$$2R\cdot \rightarrow R{-}R \quad (25)$$
  — peroxide decomposition (preventive antioxidant) $$ROOH + AH \rightarrow \text{non-radical products} \quad (26)$$

The above scheme indicates four possible ways for an antioxidant to participate in the above mechanism, i.e.

(a) *initiation* by the direct attack of oxygen on the antioxidant resulting in the formation of chain-initiating free radicals;
(b) *chain transfer* to antioxidant, where the resulting free radical reacts in given conditions, in the presence of oxygen, and restores the propagating free radicals;
(c) *termination* by hydrogen donation to the free peroxy-radical as the first step of a chain transfer, followed by the reaction of the free radical A· with a second free peroxy radical, thus terminating two kinetic chains per molecule of antioxidant;
(d) *initiation inhibition* by decomposition of hydroperoxides into stable products.

A complete inhibition of polymer oxidation through the addition of an antioxidant is evidently impossible; only a marked slowing down of the oxidation rate during the storage of the rubber goods can be expected. Antioxidant efficiency is defined as the time by which the beginning of the degradation process is delayed.

2. Relationships between structure and reactivity

Experiments on the oxidation behaviour of low molcular weight model compounds (squalene), natural rubber and synthetic polyisoprene showed that in polyisoprene, the two methylene groups (α and β) in allylic position must be considered as reactive centres for both the initial attack of oxygen and the hydrogen abstraction reaction.

$$-CH_2-\overset{\overset{\displaystyle CH_3}{|}}{C}=CH-CH_2-CH_2-\overset{\overset{\displaystyle CH_3}{|}}{C}=CH-CH_2- \quad (27)$$

The presence of the CH_3 substituent causes the methylene groups in the allylic position to have a different reactivity. Thus, a hydroperoxide (28) formed at the methylene group in the α-position with respect to the CH_3 substituent, undergoes an easier homolytic scission reaction yielding alkoxy radicals (29) [151–153].

$$-CH_2-\underset{}{\overset{CH_3}{\overset{|}{C}}}=CH-CH_2-\underset{\underset{OOH}{|}}{CH}-\overset{CH_3}{\overset{|}{C}}=CH-CH_2- \qquad (28)$$

$$-CH_2-\overset{CH_3}{\overset{|}{C}}=CH-CH_2-\underset{\underset{O\cdot}{|}}{CH}-\overset{CH_3}{\overset{|}{C}}=CH-CH_2- \qquad (29)$$

Changes in the physico-mechanical properties of rubber following degradation are the result of two competitive reactions: chain-breakdown and crosslinking.

Chain-breakdown produces volatile compounds [154, 155]. The main volatile products formed in the oxidation of polyisoprene by chain scission are levulinic aldehyde, methyl vinyl ketone and methacrolein, which still contain double bonds from the polyisoprene chain, formaldehyde and its oxidation product, formic acid.

The resulting free radicals, R· (reactions (15), (17) and (18)) may recombine in the presence of low concentrations of oxygen, forming either crosslinked or macromolecules with long side-chains.

The practical effect of crosslinking is hardening and an increased gel content, while scission is associated with softening, owing to the lower average molecular weight [156, 157] produced.

Apart from certain factors, such as temperature or oxygen concentration, which determine the prevalence of one or the other of the above-mentioned processes, the decisive role is played by the chemical structure of the polymer. Consequently, polyisoprene mostly undergoes breakdown reactions, while polybutadiene undergoes crosslinking reactions [158, 159].

Based on the chemical structure of polyisoprene and its reactivity towards oxidation, attempts were made to eliminate the centres which were active in oxidation reactions. The addition of halogens, hydrogen halides or mercaptans to the double bonds of polyisoprene has thus increased its oxidation stability, but at the same time caused undesired changes in its elastomeric properties [159].

3. Selection criteria for stabilizers

The selection of a stabilizer mainly depends on the following factors:

(a) Colour of the polymer; this factor determines the use of amine or phenolic type stabilizers. All amine type stabilizers usually produce a more or less deep staining of the polymer. Non-staining stabilizers are phenols, esters of phosphorous acid, etc.

(b) Physical properties of stabilizers. In the order of their importance they are [141, 160]:

— the physical state of the stabilizer, which necessitates particular method of metering;
— the compatibility with the polymer;
— its resistance to detergents and solvents; if they remove it, the polymer is left unprotected;
— volatility, which expresses its physical loss, for instance, during the drying of rubber, etc.;
— toxicity;
— cost/efficiency ratio.

Toxicity lately tends to occupy a more important place in the selection of a stabilizer [151, 161]. The problems related to stabilizer toxicity have orientated research towards the synthesis of non-toxic stabilizers [162].

(c) The incorporation of the stabilizers into polymer:

— into the crude polymer;
— into the cured polymer.

The stabilizers are added to the crude polymer immediately after its synthesis in low amounts of about 1–1.5 % in order to preserve the physico-mechanical properties of the polymer during the storage period before processing.

Antioxidants for cured rubber are added during compounding before vulcanization in amounts of 1–3 %, which ensures protection of the vulcanizate during both storage and weathering.

The most important classes of antioxidant are secondary diarylamines and sterically-hindered phenols, acting as kinetic chain terminating agents [141, 160]. The amine-type antioxidants are more efficient than the phenolic ones, although they stain the rubber.

The stabilization of a polymer is a complex problem. No general conclusions can be drawn on the stabilization of polymers, the data gathered for a particular polymer are not valid for another polymer. Stabilizers efficient for polyisoprene do not show the same effectiveness when added to polybutadiene, although both polymers are quite similar and belong to the same class of diene rubbers. A study of each polymer in connection with the end-use of the fabricated articles is essential.

4. Mixtures of stabilizers

As was previously discussed, polymer degradation in general and of synthetic polyisoprene in particular, as initiated and accelerated by oxygen, ozone, metallic impurities, free radicals, light, etc., is a very complex process. Therefore, just one compound—a goal still to be reached—cannot simultaneously play the role of a free radical scavenger, a metal-chelating agent and a photostabilizer. In practice, the ideal compound is replaced by a mixture of stabilizers in which the one which ensures stability under the most important stress conditions prevails [163].

In Section 2 mention was made of the fact that an antioxidant reacts with oxygen and results in the formation of free radicals capable of initiating a free radical chain oxidation (pro-oxidant effect). Research showed that each given antioxidant had an optimum concentration at which its efficiency was maximum [150]. As expected, concentrations below optimum values did not yield the desired results, while higher values give undesired pro-oxidant effects. In the case of a mixture of stabilizers, the resulting effect can take three different forms [141]:

—additive effect,
—antagonistic effect (antagonism),
—synergistic effect (synergism).

The combination of the same type of stabilizers is expected to provide only an additive effect [141]. Therefore, two stabilizers from the same class can be used, each of them at a lower concentration than that which generates a pro-oxidant effect, with the maximum efficiency to be reached by the additive effect.

The antagonistic effect is defined as a decrease in the effectiveness of the protecting system against oxidation, as a result of interactions between the stabilizers used in the mixture or between the stabilizers and the additives used in processing. Hence, the importance of testing each stabilizer mixture for the given polymer under service conditions.

The synergistic effect is apparent in those particular stabilizer mixtures which show a higher efficiency than the individual stabilizers.

Two types of synergism are differentiated [141]:

—*homosynergism*, involving two stabilizers with non-equal activities, but acting via the same mechanism;
—*heterosynergism*, involving two stabilizers acting via different mechanisms.

5. Estimation of the effectiveness of stabilizers

From the data in the literature, it is known that a stabilizer with the highest effectiveness for a given crude polymer can have a low efficiency in vulcanizates. For these reasons, the stabilizer performance in both the crude and the vulcanized polymers has been determined.

(*a*) *Estimation of stabilizer effectiveness in crude polymer*

As previously shown, the role of stabilizers added to a polymer after its synthesis is to preserve its physico-chemical and physico-mechanical properties.

Based on the fact that the selection of a given stabilizer according to its property of staining or not staining the polymer is an important factor, the following stabilizers were tested: *N*-phenyl-*N'*-cyclohexyl-*p*-phenylenediamine (4010), *N*-phenyl-*N'*-iospropyl-*p*-phenylenediamine (Santoflex IP), which are staining stabilizers, and 4-methyl-2,6-di-*tert*-butylphenol (Topanol OC), a non-staining stabilizer, at concentrations of 1 and 2% by weight based on the polymer, respectively. The effectiveness of the above stabilizers was estimated by measuring the variation with time of certain polymer characteristics, such as Mooney viscosity, the intrinsic viscosity of the soluble fraction and the gel content (Table 16).

From the data shown in Table 16, we concluded that any of the abovementioned stabilizers can be used to preserve the physico-chemical and physico-mechanical properties of polyisoprene during the storage period up to its processing. No differences in efficiency were noticed. The selection of a stabilizer is therefore based on the intended use of the polymer (in both coloured or non-coloured rubber goods) and on the cost/efficiency ratio.

During storage, when a polymer is not subject to severe temperature and pressure conditions, the only process is shelf ageing which occurs at a slow rate in stabilized polymers. Therefore, the changes in the physico-mechanical properties of a polymer cannot be used to estimate the effectiveness of a stabilizer. Accelerated ageing tests which simulate the operating conditions of the rubber goods have to be carried out, instead.

One of the most widely used methods in estimating the effectiveness of stabilizers added to polymers under accelerated ageing conditions is the oxygen absorption of the polymer samples. Oxygen-absorption tests can be made either under isobaric conditions (the volumetric method—which measures the decrease in oxygen volume at constant pressure [136, 164–166]) or isochoric conditions (the manometric method—which measures the decrease in oxygen pressure at a constant volume [146]).

A special device was constructed to enable oxygen-absorption determinations to be made by volumetric measurements [167]. Oxygen

Table 16

Effect of the nature and amount of stabilizer on polyisoprene properties stored for 1 year in normal conditions

(initial polyisoprene: $[\eta] = 4.5$ dl/g; gel content 20%; Mooney viscosity ML $(1 + 4)$ 100°C = 72)

	Stabilizer		Polyisoprene characteristics		
No.	Type	Concentration, g/100 g polyisoprene	$[\eta]$, dl/g	Total gel, %	Mooney viscosity (ML$(1 + 4)$ 100°C
1.	Santoflex IP	1	4.45	23	73
2.	Santoflex IP	2	4.45	23	72
3.	4010	1	4.38	22	71
4.	4010	2	4.30	22	72
5.	Topanol OC	1	4.30	23	70
6.	Topanol OC	2	4.30	23	70

absorption by the polymer sample was made at atmosphere pressure and at temperatures chosen *a priori* and kept constant.

The device was used to determine the absorption of oxygen by thin films of natural rubber, synthetic polyisoprene and polybutadiene rubbers. Reaction isotherms, determined with the above device, helped us to evaluate the effectiveness of certain stabilizers for synthetic polyisoprene. The induction periods ($\tau_{140\,^{\circ}C}$) of the oxidation reaction of synthetic polyisoprene containing phenolic or amine-type stabilizers are listed in Table 17.

Data shown in Table 17 indicate that from the point of view of protecting the polymer against the combined action of oxygen and temperature, the most efficient are the amine-type stabilizers, i.e. Santoflex IP and 4010.

Table 17

Induction period of the oxidation reaction of synthetic polyisoprene containing phenolic and amine-type stabilizers

No.	Stabilizer	g/100 g polyisoprene	$\tau_{140\,^{\circ}C}$ (min)
1	Santoflex IP	1	710
2	4010	1	600
3	Topanol OC	1	40
4	Topanol OC	2	86

(*b*) *Estimation of the effectiveness of antioxidants in vulcanized rubber*

In order to estimate the effectiveness of an antioxidant in vulcanizates, recipes based on *cis*-1,4-polyisoprene (Table 18) were made up. Their physico-mechanical properties are given in Table 19.

The resulting test specimens were subject to the following accelerated and shelf-ageing tests:

— oven ageing at 100°, for 1, 3 and 5 days in air (Table 20);
— ageing at 100°, under a pressure of 10 atm oxygen for 24 hours (Table 21);
— ageing in ozone atmosphere (50 ppm), at 25°C and under 20 % extension, for 103 hours (Table 22);
— static shelf ageing for 6 weeks to determine the variation in the physico-chemical and physico-mechanical properties, measured at 15-day intervals.

In the accelerated ageing tests on cured rubber, only amine-type antioxidants were used, i.e. Santoflex IP and antioxidant H (2,2,4-trimethyl-1,2-dihydroquinoline).

Table 18
Experimental stocks (parts by weight)

Compounds	Batch 0	Batch 1	Batch 2
Polyisoprene rubber	100	100	100
HAF black	50	50	50
Stearic acid	1	1	1
Zinc oxide	5	5	5
Oil	3	3	3
Santoflex IP	—	2	—
Flechtol H	—	—	2
Vulcacit CZ (CBS)	0.9	0.9	0.9
Sulphur	2.8	2.8	2.8

Table 19
Physico-mechanical properties of experimental stocks based on *cis*-1,4-polyisoprene*

Property	Unit	Batch 0	Batch 1	Batch 2
Curing time	min	10	7	10
300% modulus	kgf/cm^2	105	98	102
Tensile strength	kgf/cm^2	274	278	273
Elongation at break	%	530	523	480
Set at break	%	20	20	16
Shore hardness	°Sh	63	66	63

* Polyisoprene used had ML (1 + 4) at 100°C of 66.

The results of the static shelf-ageing tests are shown in Figure 31.

The variation of the physico-mechanical properties of the *cis*-1,4-polyisoprene vulcanizates subject to the above accelerated ageing tests leads to the following observations:

— the efficiency of the two tested antioxidants is similar in both the thermal oxidation and oxygen pressure ageing tests, while in the ozone and shelf ageing tests, Santoflex IP has a markedly higher efficiency. Hence, the conclusion is that apart from a good antioxidant effect, Santoflex IP also confers protection against ozone;

— the antioxidant and antiozonant properties of Santoflex IP recommend its use in the compounding of vulcanizates while the use of antioxidant H undoubtedly needs the addition of an antiozonant as well.

6. Macromolecular antioxidants

For a given stabilized polymer, loss of antioxidant is likely to occur not only by its consumption in oxidation prevention [160]. In fact, physical losses of

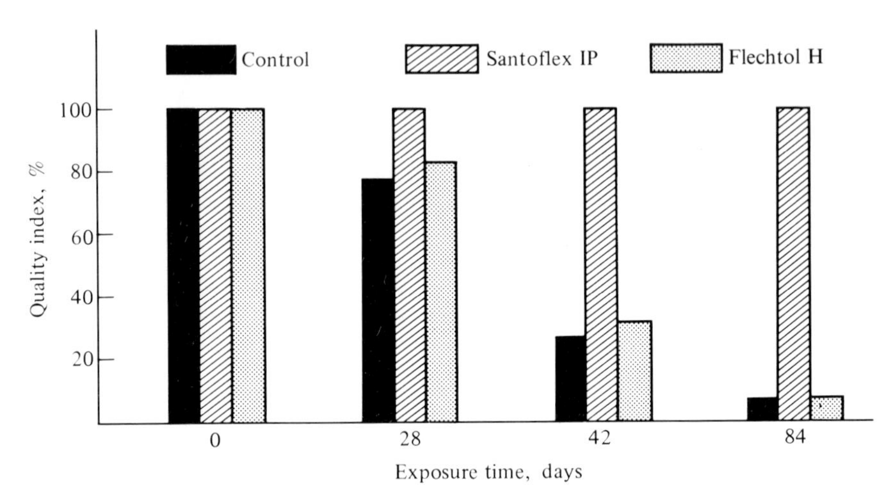

FIGURE 31. Variation of quality index of cured stocks based on *cis*-1,4-polyisoprene as a function of weathering time.

Table 20
Thermal oxidative ageing of cured stocks based on *cis*-1,4-polyisoprene

Property	Unit	Batch 0	Batch 1	Batch 2
Tensile strength	kgf/cm^2			
Normal		274	278	273
Aged 1 day		183	219	214
3 days		131	170	161
5 days		99	126	121
Loss in tensile strength	%			
1 day		33.5	21.2	21.5
3 days		52.2	38.5	40.8
5 days		64.2	54.1	55.5
Elongation at break	%			
Normal		530	523	480
Aged 1 day		305	300	330
3 days		200	210	320
5 days		137	157	170
Loss in elongation at break	%			
1 day		42.4	42.6	20.8
3 days		62.2	59.8	33.3
5 days		74.1	69.9	64.5

antioxidants take place; in certain conditions such losses become very important. They are produced by:

— antioxidant volatility at high temperatures, low pressures or when the rubber goods are exposed to steam, which can increase the volatility by steam distillation;
— washing with water and detergents;
— extraction with solvents.

Table 21
Oxygen pressure ageing of cured stocks based on *cis*-1,4-polyisoprene

Property	Unit	Batch 0	Batch 1	Batch 2
Tensile strength	kgf/cm^2			
Normal		274	278	273
Aged		96	130	158
Loss in tensile strength	%	65.1	53.2	42.1
Elongation at break	%			
Normal		530	523	480
Aged		283	320	330
Loss in elongation at break	%	46.7	38.3	31.2

Table 22
Resistance to ozone cracking of cured stocks based on *cis*-1,4-polyisoprene

Time (hr)	Batch 0 a	Batch 0 b	Batch 1 a	Batch 1 b	Batch 2 a	Batch 2 b
1	4	1	—	—	4	$\frac{1}{2}$
2	4	4	—	—	4	$\frac{1}{2}$
3	4	4	—	—	4	$\frac{1}{2}$
4	4	4	—	—	4	$\frac{1}{2}$
6	4	4	—	—	4	1
7	4	4	4	$\frac{1}{2}$	4	1
19	4	4	4	$\frac{1}{2}$	4	1
55	4	4	4	1	4	1
79	4	7	4	1	4	1
103	4	7	4	1	4	1

Note: The figures in column a correspond to the density of surface cracking: 1 = 1/8 of the surface; 2 = 1/8–3/8 of the surface; 3 = 3/4 of the surface; 4 = >3/4 of the surface. The figures in column b are estimates of crack length: 1/2 = 0.2 mm; 1 = 0.2–0.4 mm; 4 = 0.8 mm; 7 = 2 mm; 10 = 5–6 mm.

In service conditions such physical losses produce a shortening of the life of the rubber goods.

In order to reduce antioxidant volatility and solubility, higher molecular-weight compounds were used based on the known volatility—mobility—molecular-weight relationship. Therefore, our research on synthesis of an antioxidant focused on macromolecular antioxidants.

The advantages of macromolecular antioxidants as compared to the low molecular ones are the following [152, 153, 168–175]:

— they are not volatile at high temperatures and low pressures;
— they have a higher effectiveness in a wider polymer range;
— they have good compatibility with polymers, especially those with a closely related chemical structure;
— there is no or little interference with curing agents;
— they are stable to solvent extractions.

The effectiveness of macromolecular antioxidants synthesized by us was tested [167] and compared with the low molecular *N*-phenyl-*N'*-isopropyl-*p*-phenylenediamine antioxidant (Santoflex IP) [161]. Selection of Santoflex IP was based on the previously-reported data which indicate it to be one of the most efficient low molecular antioxidants.

As the large differences between the molecular weights of the low molecular and macromolecular antioxidants render the use of percent concentration for the tested antioxidants incorrect, we worked at constant nitrogen concentrations assuming that each N—H bond in both the low molecular and the macromolecular antioxidants can be involved in trapping free radicals, thus inhibiting the thermal oxidative degradation of synthetic polyisoprene.

A concentration of 0.0248 g nitrogen equivalent to 0.2 g Santoflex IP per 100 g *cis*-1,4-polyisoprene was arbitrarily chosen for testing.

The tested macromolecular antioxidants, marked AOM_i ($i = 1$–5), contain nitrogen in various proportions and have number-average molecular weights ranging from 2000 to 12,000.

Effectiveness of macromolecular antioxidants was estimated from the induction periods, $\tau_{140\,^\circ C}$, in thermal oxidation.

In order to outline the above-mentioned characteristics of macromolecular antioxidants, the rubber samples containing the tested antioxidants were extracted with solvents (toluene–ethanol azeotropic mixture, b.p. = 76.65°C) in a Sohxlet and tested again for oxidation stability.

From data listed in Table 23, the following observations can be made:

— all the macromolecular antioxidants which were tested inhibited the thermal oxidation of synthetic polyisoprene;
— the maximum induction period, $\tau_{140\,^\circ C}$, for macromolecular antioxidants was given by the one with the lowest number-average molecular weight.

Table 23

Induction period ($\tau_{140^\circ C}$) of the oxidation reaction of stabilized synthetic polyisoprene before and after extraction

No.	Antioxidant	Nitrogen (%)	$\overline{M}_n$	$\tau_{140^\circ C}$ (min) Before extraction	After extraction
1	AOM-1	0.785	2000	160	96
2	AOM-2	0.505	3400	134	90
3	AOM-3	0.242	5000	94	87
4	AOM-4	0.164	7500	86	71
5	AOM-5	0.114	11,700	76	70
6	Reference Santoflex IP	12,389	226	170	9
7	Polyisoprene without antioxidant	—	400,000	10	

As the molecular weight increases, the macromolecular antioxidant efficiency decreases. This efficiency decrease must be related to a reduction in mobility as the number-average molecular weight increases. Although the advent of macromolecular antioxidants invalidates the postulates [159, 160, 176] which assumed that antioxidant effectiveness is mainly determined by its mobility [153, 154, 170] (mention should be made that no unity of views on the mobility-effectiveness relationship so far exists), it could be expected that as mobility decreases, antioxidant effectiveness will also be reduced. The reduction of antioxidant mobility implies their more difficult extraction from the polymer, as also shown in Table 23. The induction period before and after extraction remains virtually unchanged for macromolecular antioxidants with higher number-average molecular weights. As to the antioxidants with lower number-average molecular weights, the induction periods decrease, but still remain within suitable limits, much higher than those determined for polymers originally stabilized with Santoflex IP, which is completely extracted by solvents. The oxygen-absorption isotherm of the sample containing Santoflex IP after extraction becomes almost identical with that of the polyisoprene sample without antioxidant.

In order to reduce the amount of macromolecular antioxidants added to rubber, mixtures of low molecular and macromolecular antioxidants were tested. Thus, we tested a mixture of AOM-2 and Santoflex IP, keeping constant an arbitrarily chosen level of nitrogen of 0.0248 g per 100 g *cis*-1,4-polyisoprene.

Table 24

Variation of the induction period of the oxidation reaction of *cis*-1,4-polyisoprene as a function of the composition of Santoflex IP and AOM-2 antioxidant mixture

No.	Antioxidant mixture		g antioxidants/100 g rubber	$\tau_{140°C}$ (min) Before extraction	After extraction
1	Santoflex IP AOM-2	100% 0%	0.20	170	9
2	Santoflex IP AOM-2	70% 30%	1.61	136	30
3	Santoflex IP AOM-2	50% 50%	2.55	182	135
4	Santoflex IP AOM	30% 70%	3.49	116	74
5	Santoflex IP AOM-2	0% 10%	5.51	134	100

From the data shown in Table 24 and in Figure 32, it is seen that a mixture of 50% of low molecular and macromolecular antioxidants resulted in maximum induction periods, both before and after extraction.

The advantage of using mixtures of low molecular and macromolecular antioxidants arises from the fact that unlike Santoflex IP, the macromolecular antioxidants do not confer protection against ozone too.

The complexity of the polymer-degradation process explains the extended research in this field and the ever-growing number of patents describing

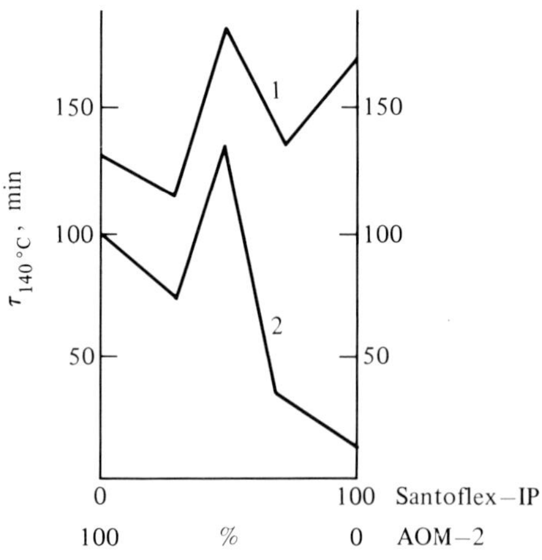

FIGURE 32. Variation of the induction period in the oxidation of *cis*-1,4-polyisoprene as a function of the composition of the Santoflex IP and AOM-2 antioxidant mixture. 1. Before extraction; 2. After extraction.

syntheses of stabilizers which ensure polymer protection in most varied working conditions.

In recent years, intensive studies on accelerated controlled polymer degradation have been made at the same time as investigations on polymer stabilization. Such experiments were prompted by the need to find a solution to the problem of polymer wastes which accumulate every year and are now a really serious ecological problem. The ideal would be if stabilizers used in polymer protection allowed a conservation of all properties of rubber goods for precalculated service lives and that beyond this their action stopped suddenly, producing a controlled degradation of the polymeric material.

Chapter V

Recovery of polyisoprene from solution

THE use of stereospecific organometallic catalysts in the synthesis of high molecular-weight stereoregular polymers results in widely used industrial processes based on methods of solution polymerization in organic solvents. A highly viscous solution is obtained at the end of the polymerization reaction, with a relatively low content in polymer: this is because of the high molecular weight required by the physico-mechanical characteristics needed for the use of polyisoprene in various fields. Therefore, attempts were made to find a method of separation of the polymer from solution and a method to recover in large amounts the solvent present.

In the separation of a polymer from solution, a number of questions arises, since the physico-chemical and physico-mechanical characteristics of the final products largely depend on the polymerization processes and on the working conditions during this stage.

On the other hand, one must take into account that the method used in polymer and solvent recovery has a major role in the economics of the industrial process.

The main goal of polymer separation from solution is an almost complete removal of solvent in order to avoid the risk of blister formation in the rubber during vulcanization. At the same time, separation must be carried out in such conditions as not to be detrimental to the properties of the rubber, i.e. so as not to produce polymer degradation and crosslinking. A great importance is also attached to the economic aspects related to the consumption of steam and energy. This is the most energy-consuming stage in all the manufacturing processes for stereoregular rubbers [93].

The known methods for polymer separation from solution are the precipitation of the polymer with a non-solvent in the absence of water [177] or solvent removal in the presence of water [178–182].

The separation methods in the absence of water involve polymer precipitation with corresponding non-solvents, i.e. alcohols and ketones, and at first sight show certain economic advantages, when compared to the methods using water, since tedious drying operations of the recovered solvent and monomer are avoided. The method, however, has certain disadvantages of its own [93]:

— The precipitating agents, namely alcohols and ketones, which are at the same time catalyst deactivating agents, form, by reaction with the catalyst, products which remain in the polymer and are very difficult to remove. Such products favour polymer degradation.
— Polymers precipitated from their solutions are strongly swollen, viscous and sticky materials and include large amounts of solvent, difficult to remove.
— One has to consider problems related to the fractionation of the precipitating agent–solvent mixture, as well as the fact that a continuous and economical transport of the separated polymer cannot be achieved.

The above-mentioned disadvantages limit the application of such methods on an industrial scale.

Polymer separation from solution with warm water or with steam is the basic industrial method and has advantages related to the achievement of a continuous system, an efficient removal of solvent and the possibility of transport through pipes of the aqueous polymer slurry. Several polymer-recovery processes are reported in the literature, based on treatment of the polymer solution with water and/or steam, which are non-toxic and cheap utilities. Solvent is removed with steam and the polymer remains in the form of particles in water. This very important step in polyisoprene manufacturing involves dispersion of the polyisoprene solution in an aqueous medium at a temperature high enough for solvent evaporation to be achieved. For the operation to be successful solvent removal must be complete and the polymer particles must not be agglomerated.

In order to reach optimum conditions, two basic problems must be solved:

— formation of droplets of a suitable size;
— prevention of particle agglomeration by fusion.

In general, control of droplet sizes and prevention of particle agglomeration depend on the presence in the system of a suspension stabilizer, on its nature and concentration, on the water/organic phase ratio, pH, stirring efficiency and temperature.

The suspension stabilizer, also called the *dispersing agent*, produces an increase in the viscosity of the aqueous phase, which favours an efficient dispersion of the organic phase in the aqueous phase. The dispersing agents, with surface active properties, also produce a decrease in the surface tension of the water. They result in the formation of rubber particles of small size which permit good heat and mass transfer, as well as allowing solvent to diffuse through the rubber particles into the aqueous layer. The surfactant coats the surface of each particle with a thin film preventing a direct contact between the particles so that further agglomeration is avoided.

A good dispersing agent produces small and uniform particles. In selecting a dispersing agent, one should not forget that its efficiency varies even within the same class of substances when referring to a macromolecular dispersing agent. The efficiency of a dispersing agent is thus dependent on the degree of polymerization, the degrees of esterification and hydrolysis, and its solubility in the aqueous and organic phases, etc.

The particle size is reduced as the concentration of the dispersing agent increases. A particular dispersing agent concentration affording particles with an optimum size should therefore be found. The optimum concentration of dispersing agent depends in its turn on the phase ratio, the temperature, the pH of the medium and the stirring.

High aqueous/organic phase ratios favour the formation of particles with small and uniform sizes and in this way heat and mass transfer are improved. In this case too, optimum conditions must, however, be found, because too high a ratio is uneconomical as it involves the handling of too much water.

The stability of the suspension is favoured by a neutral or slightly alkaline pH.

An efficient stirring at high peripheral speeds gives particles of small size and affords a stable suspension, as well as producing improved transfer processes. In order to obtain particles with convenient sizes, efficient stirring can be associated with feeding the polymer solution into the aqueous phase through special nozzles, the so-called particle-forming nozzles, which ensure good dispersion of the polymer solution.

Temperature depends on the nature of the polymerization solvent and should be higher than the boiling point of the solvent–water azeotrope. In practice, solvent removal from the polymer-recovery vessel at a given temperature is limited by the equilibrium between the partial pressures of water and solvent vapours. High temperatures have an undesired effect upon suspension stability. Therefore, one must choose a dispersing agent and working conditions such as dispersing agent concentration, pH, organic/aqueous phase ratio, where the suspension is stable and optimum particle sizes are reached.

Polymer recovery under reduced pressure provides a more efficient removal of solvent. This technique is recommended for the removal of solvents with high boiling points or for separating polymers which are unstable at higher temperatures.

The following types of compounds can be used as dispersing agents:

- —natural macromolecular compounds such as soluble starch, gelatin, pectin, or dextrin;
- —soluble synthetic macromolecular compounds or cellulose derivatives such as poly(vinyl alcohol), poly(ethylene oxide), the copolymer of vinyl

acetate and maleic anhydride, the copolymer of styrene and maleic anhydride, methyl cellulose and carboxymethyl cellulose;
— inorganic salts of some fatty acids, such as zinc stearate;
— inorganic powders, such as talc, kaolin, barytes, kieselguhr, or calcium phosphate.

As the use of inorganic salts greatly increases the ash content of the polymer, organic dispersing agents are chosen which, apart from their action as dispersing agents during the polymer recovery step, can also assist in dispersing reinforcing fillers such as carbon black [183]. Such types include organic dispersing agents which can be non-ionic, anionic or cationic, all of them containing in their molecule a water-soluble part and a part soluble in organic solvents. Two dispersing agents can be used simultaneously, i.e. a mixture formed from an anionic surfactant (anionic electrolyte) and a cationic surfactant (cationic electrolyte). Anionic electrolytes can be salts of fatty acids, sulphonated higher alcohols, etc., while cationic electrolytes are aminoaliphatic salts, quaternary ammonium salts, etc. Some macromolecular compounds with anionic or cationic character can also play the role of surface active agents. For instance, polymers of acrylic acid or methacrylic acid or their salts, copolymers of dibasic unsaturated acids such as maleic or fumaric acids, macromolecular compounds with sulphonic groups, all have an anionic character; polysoaps of polyvinylpyridine type, copolymers of polyacrylamide, polyacrylic esters, etc., all have a cationic character. Such mixed surface-active agents are used in the proportion of 0.1 to 2 parts by weight per 100 parts of polymer [183].

Also, cellulose derivatives containing both hydrophylic and hydrophobic groups in their molecule can be used such as: methyl cellulose, carboxymethyl cellulose, or hydroxymethyl cellulose [184].

The dispersing agents can be introduced as a solution into either water or the polymer solution.

The average diameter of the polymer particles depends on the nature of the dispersing agent, on the stirring system, on the polymer solution feed rate as well as on the viscosity of the polymer solution [184, 185].

In experiments on the separation of polymer from solution, both basic procedures were tested, i.e. precipitation with a non-solvent and desolventation.

Polymer precipitation from solution was achieved with a mixture of a solvent and a non-solvent, using the following liquids: heptane–methanol, heptane–ethanol and heptane–acetone.

By using this method, the resulting polyisoprene contains a considerable amount of ash as the result of insufficient removal of catalyst residues imbedded into the polymer, which further favour rubber degradation. The lower the molecular weight of the polyisoprene, the higher the ash content as the result of polymer stickiness, which prevents good washing. Moreover, polymer separation by precipitation needs very large quantities of solvent and non-solvent, which render the method uneconomical.

In polymer separation from solution in the presence of water, the recovery of the polymer was totally separated from the stages of catalyst system deactivation, polymer stabilization and catalyst residue washing. The investigations were aimed at working out a manufacturing process and a recovery unit where a non-agglomerated, finely divided, completely solventless final product would be obtained, easy to handle and to convey the further stages of filtration and drying, while maintaining the physico-chemical and physico-mechanical characteristics of the polymer [186, 187].

Special attention was paid to the study of the nature of the dispersing agent as well as to the concentration in which it must be introduced during the separation stage. The lower the size of the polymer particles after recovery, the higher the specific surface and hence the more efficient the solvent removal. On the other hand, the rubber particles should not be too small, as they must be retained on the filter. Good results were obtained with a water-soluble, macromolecular, cellulose derivative, i.e. methyl cellulose.

In order to extend the number of dispersing agents, other compounds of the cellulose class were tested, namely carboxymethyl cellulose and hydroxymethyl cellulose as well as two types of poly(vinyl alcohol) with two different degrees of hydrolysis. The physico-chemical properties of those compounds are listed in Tables 25 and 26.

The data obtained with these dispersing agents are shown in Table 27. From this, one can see that by working with a polymer solution in isopentane, where the various dispersing agents were introduced as aqueous solutions from the beginning, fine rubber particles were obtained. The size of those particles depends on the nature of the dispersing agent and on its concentration with respect to the amount of polymer present in solution.

In order to establish the influence of dispersing agent concentration, polymer separations at various ratios of dispersing agent and polymer were also carried out.

At the same dispersing agent/polymer ratio, it was found that the finest particles formed after removal of the solvent were obtained in the presence of methyl cellulose. The other dispersing agents also gave satisfactory results.

It was assumed that the determination of the residual solvent in polymer after recovery could serve as a method of estimating the efficiency of the dispersing agent. Therefore, the total volatile content (the sum of solvent plus water) of the polymer after recovery was determined. Water was extracted from the polymer with a mixture of methanol and toluene (7:3 v/v) and was determined with Karl Fischer reagent. The residual amount of solvent was then calculated by difference. The results of such determinations are given in Table 28.

The lowest amounts of solvent in the polymers were obtained when methyl

Table 25

Physico-chemical characteristics of cellulose derivatives

No.	Compound	Content of active substance (%)	Degree of polymerization	Ether groups/ structural unit	Solution[a] viscosity (cP)	Moisture (%)
1	Methyl cellulose	93	500–1000	1.5	21–26	7.0
2	Carboxymethyl cellulose	92	—	0.663	16.8	7.0
3	Hydroxyethyl cellulose	—	550	1.00	400	—

[a] Measured in 2% aqueous solutions at 20°C.

Table 26
Physical-chemical properties of poly(vinyl alcohol) (PVA)

No.	Degree of polymerization	Degree of hydrolysis (%)	Hydrolysis number mg KOH/g PVA	Solution viscosity[a] (cP)
1	1100	86.4	153.3	—
2	1100	89.9	117.2	11.66

[a] Measured in 4% aqueous solutions at 20°C.

Table 27
Desolventation conditions and qualitative results

No.	Dispersing agent	Concentration g/100 g polymer	Remarks
1	Methyl cellulose	1.0	Fine particles
2	Methyl cellulose	2.0	Very fine particles
3	Carboxymethyl cellulose	2.0	Agglomerated particles attached to the stirrer paddles
4	Carboxymethyl cellulose	4.0	Particles 0.5–1.0 cm. Good desolventation
5	Hydroxyethyl cellulose	2.0; 4.0	0.5–1.5 cm particles
6	Poly(vinyl alcohol) (degree of hydrolysis = 86.4%)	2.0; 1.0	0.5 cm particles, without agglomeration
7	Poly(vinyl alcohol) (degree of hydrolysis = 89.9%)	2.0; 4.0	Good desolventation, but later agglomerated

pH of the desolventation solution = 7–8; temperature = 55°C; concentration of the aqueous solution of the dispersing agent = 5%.

cellulose and poly(vinyl alcohol) with a degree of hydrolysis of 86.4% were used as dispersing agents.

From data in the literature it is known that among the various types of poly(vinyl alcohol), the one with a degree of hydrolysis between 80 and 87% [188] yields the best results. Our experimental results are in agreement with those findings.

The influence of the conditions of recovery on the characteristics of the rubber, namely the molecular weight, and polymer gel and ash content of the polymer, was also investigated.

As shown from the data listed in Table 29, the use of the above described process of separation has little effect on the properties of the polymer.

Table 28

Effect of dispersing agents on polyisoprene recovery

No.	Dispersing agent			Volatiles in polyisoprene, %		
	Type	Degree of hydrolysis, %	concentration, g/100g polymer	Water	Solvent	Total
1	Methyl cellulose	—	2	34.56	0.5	35.06
2	Poly(vinyl alcohol)	86.4	2	27.88	0.5	28.38
3	Poly(vinyl alcohol)	86.4	2	23.46	0.6	24.06
4	Poly(vinyl alcohol)	89.9	4	30.23	2.0	32.23
5	Hydroxyethyl cellulose	—	4	31.00	1.67	32.67
6	Carboxymethyl cellulose	—	4	23.81	1.1	24.91

Table 29

Effect of desolventation on intrinsic viscosity, gel content and ash content of polyisoprene[a]

No.	$[\eta]$ (dl/g)		Gel content (%)		Ash content after desolventation (%)
	Before desolventation	After desolventation	Before desolventation	After desolventation	
1	3.48	3.30	13.6	16	0.19
2	3.58	3.35	10.0	12.5	0.20
3	3.38	3.25	12.6	10	0.18
4	3.36	3.22	12.4	11.4	0.18

[a] Polyisoprene solution in isopentane; dispersing agent = methyl cellulose; 2 g/100 g polymer.

Chapter VI

Characterization of *cis*-1,4-polyisoprene rubber by nuclear magnetic resonance, electron microscopy and electron diffraction

NEW and more accurate methods of investigation of the physico-chemical properties of *cis*-1,4-polyisoprene rubber, such as nuclear magnetic resonance (NMR), electron spin resonance (ESR) and electron microscopy, have been developed in parallel with the improvement of the manufacturing processes.

1. Determination of the microstructure of *cis*-1,4-polyisoprene by high-resolution nuclear magnetic resonance

Nuclear magnetic resonance and infrared spectroscopy (a method with a variability of $\pm 3\%$, as given by the most modern instruments) are now used to determine the microstructure of isoprene polymers. If a high resolution instrument is used, the method permits a determination of the microstructure with a variability below $\pm 1\%$. To estimate the efficiency of the new polymerization catalyst systems, such a method was needed since an increase or a decrease by a few units in the *cis*-1,4- isomer content in polyisoprene produces a considerable change in its physico-chemical properties.

Synthetic polyisoprene contains the following four types of isomeric units: *cis*-1,4, *trans*-1,4, 1,2- and 3,4-, with the following structures:

```
 H3C       H              -H2C       H
    \     /                   \     /
     C=C                       C=C
    /     \                   /     \
-H2C       CH2-           H3C        CH2-
   cis-1,4                   trans-1,4
```

$$\begin{array}{c} CH_3 \\ | \\ -CH_2-C- \\ | \\ CH \\ \| \\ CH_2 \end{array} \qquad \begin{array}{c} -CH_2-CH- \\ | \\ C-CH_3 \\ \| \\ CH_2 \end{array}$$

1,2 3,4

The protons of such structural units generate characteristic lines in high-resolution NMR spectra. Quantitative analyses of the spectra can be made, since there is a direct correlation between the number of protons generating a particular NMR signal and the area under the absorption curve.

Based on NMR spectra, Chen [189] succeeded in determining the isomeric composition of a butadiene–isoprene copolymer identifying 1,2-, 1,4- and 3,4-units. The same author later outlined the possibility of determining the *cis*/*trans* ratio of the 1,4-structural units, by working with artificial mixtures of *cis*-1,4 and *trans*-1,4 isomers [190]. Although relatively wide, the proton resonance lines of polyisoprene solutions in carbon tetrachloride are, however, almost resolvable and can therefore be clearly assigned to the various structural units. Their identification is based on chemical shifts [189, 190].

Further studies indicated that the 1,4-isomer generates a characteristic line at $\delta = 5.2$–5.08 ppm and its methyl group has a chemical shift $\delta = 1.67$ ppm in the *cis* position and $\delta = 1.60$ ppm in the *trans* position. The 3,4-isomer shows a characteristic doublet at $\delta = 4.67$–4.73 ppm, while the 1,2-isomer has absorption lines at 5.30 and 0.92 ppm [191, 192].

The molar ratio between the 1,4- and 3,4-structural units is determined by the ratio of the areas of the 5.05- and 4.7-ppm lines, while the ratio between (*cis*-1,4 + 3,4) and the *trans*-1,4 structural units is calculated from the ratio of the areas of the lines given by the methyl groups at 1.67 and 1.60 ppm, respectively. The determination of the isomeric composition of diene polymers has been discussed by other authors, too [193–195].

Three types of polyisoprene rubbers, i.e. natural rubber (a), *cis*-1,4-polyisoprene prepared with a Ziegler–Natta-type complex catalyst (b) and *cis*-1,4-polyisoprene prepared with butyllithium (c) were analysed by nuclear magnetic resonance. The measurements were made on 3% (w/w) solutions of polyisoprene in carbon tetrachloride.

The spectra were recorded by the pulse method using a Fourier transform procedure. A Bruker SxP-4-100 type NMR spectrometer was used with a B-E-38 high-resolution magnet equipped with a BNC-12 type Fourier processor.

To stabilize the magnetic field, a capillary of 1 mm diameter filled with D_2O was used, since the spectrometer was provided with a B-SN-20 deuterium heteronuclear stabilizer.

In the first series of spectra, recorded at room temperature, the spectral region around $\delta = 4.7$ ppm was covered by the line of residual protons present in heavy water. Therefore, no information could be obtained on the 3,4-units.

The measurements were repeated at a somewhat higher temperature (about 63°C). The different chemical shift of water protons ($\delta_{63^\circ} = 4.44$, and $\delta_{30^\circ} = 4.7$ ppm) allowed the evaluation of the 3,7-unit content in the polymer.

The occurrence of *trans*-1,4- and 3,4-units was especially observed in polyisoprene prepared with butyllithium (c). In the other two samples, it seems that the polymers preponderantly contain *cis*-1,4-units (Figures 33a–c).

The results are given in Table 30.

2. Determination of glass-transition temperatures by electron spin resonance

Electron spin resonance measurements have proved to be highly important in studies on synthetic polymers and copolymers. Such studies are based on the presence of a paramagnetic centre in the sample. Data on glass-transition temperature, T_g, as well as other physical properties can be obtained from an analysis of the resonance signal of such a paramagnetic centre [196, 197].

A major purpose of our research was to determine the glass-transition temperature, T_g, of some samples of polyisoprene rubber compared with samples of polybutadiene rubber. Knowledge about this property is important as it denotes the temperature above which the polymer behaves as an elastomer and below which it behaves like a glassy material. The T_g value also

Table 30

Microstructure of natural rubber and of *cis*-1,4-polyisoprenes synthesized by various methods as determined by NMR

Rubber type	Microstructure (mole %)			
	cis-1,4	*trans*-1,4	1,2	3,4
Natural rubber	98	2	—	—
cis-1,4-Polyisoprene synthesized with Ziegler–Natta catalyst	98	2	—	—
cis-1,4-Polyisoprene synthesized with butyllithium	84.9	9.9	—	5.2

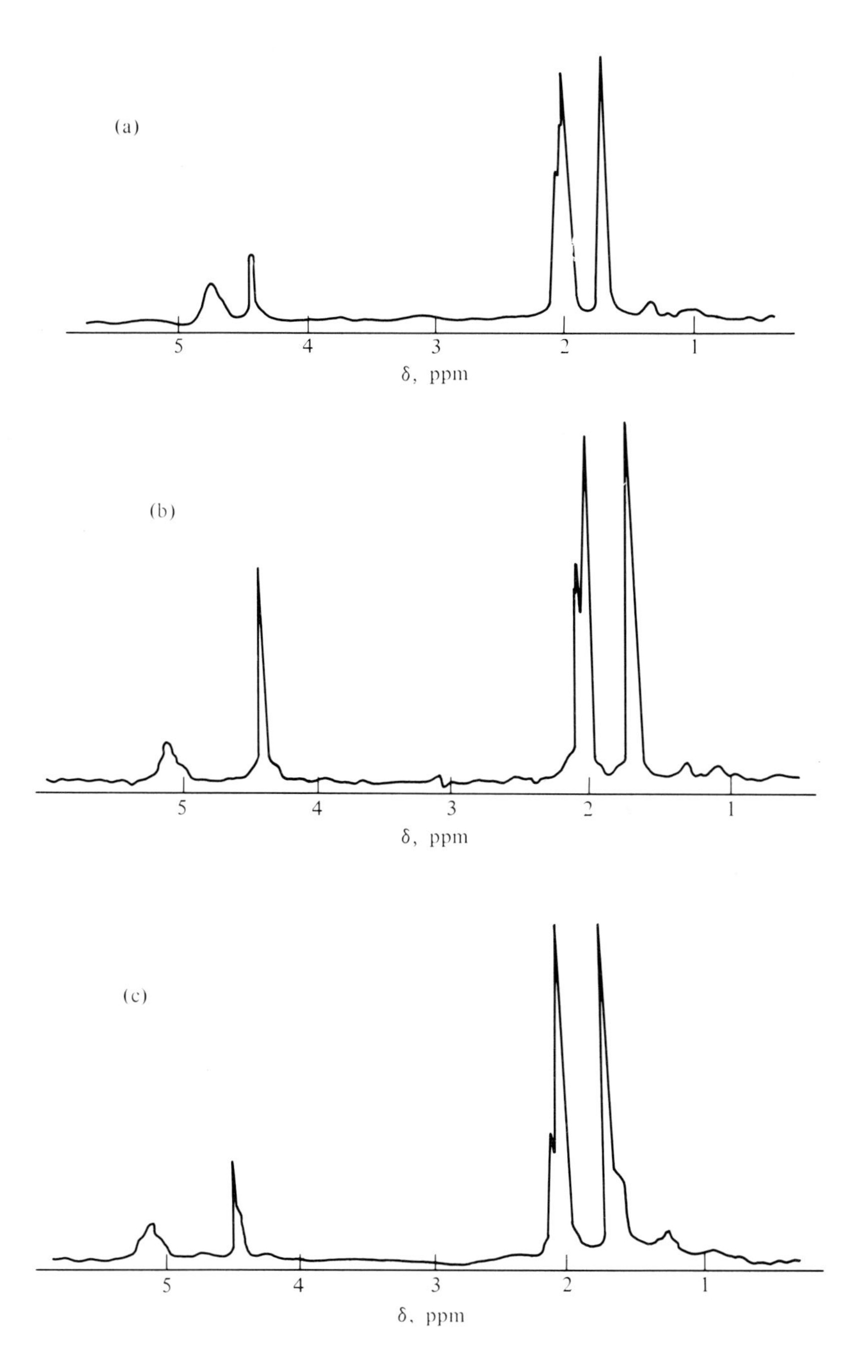

FIGURE 33. NMR spectra of natural rubber (a) and of *cis*-1,4-polyisoprene synthesized with a Ziegler-Natta catalytic complex (b) and with butyl-lithium (c).

provides information on the microstructure and molecular weight distribution of elastomers.

According to the way the paramagnetic centres are incorporated in the polymer matrix, two types of electron spin resonance experiments can be devised:

(a) *spin probe* experiments, where *free* paramagnetic molecules are present within the polymer matrix in concentrations ranging from 10 to 100 ppm;

(b) *spin labelling* experiments, where the paramagnetic centre is included in the polymer backbone.

In our experiments on determinations of glass-transition temperatures, we used the spin probe technique described by Kumler and Boyer [198].

Measurements are usually carried out in the X-band (9 GHz). T_g values can be calculated from the variation with temperature of the width of the resonance line. By plotting these values, curves similar to those shown in Figure 34 are obtained [197]. The T_{50G} value is determined from such curves, i.e. the temperature corresponding to the reduction of the line width to half of its value in the $T < T_g$ and $T > T_g$ range, respectively. The T_{50G} parameter can be further correlated with both the glass-transition temperature and the melting temperature of the polymer.

The temperature dependence of the line width for the *cis*-1,4-polybutadiene and *cis*-1,4-polyisoprene synthesized by us is shown in Figure 35.

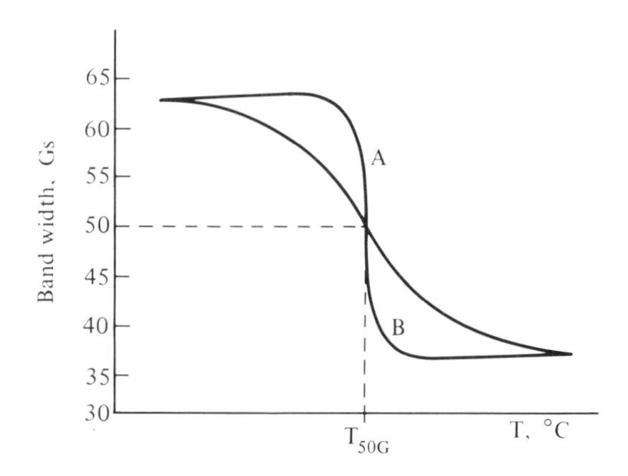

FIGURE 34. Variation of ESR line width with temperature. A—ideal transition; B—distorted transition.

As shown in Figure 35, $T_{50G} = -4.8°C$ was obtained for synthetic *cis*-1,4-polyisoprene and $-50°C$ for *cis*-1,4-polybutadiene. These values are further correlated with the glass-transition temperatures, T_g.

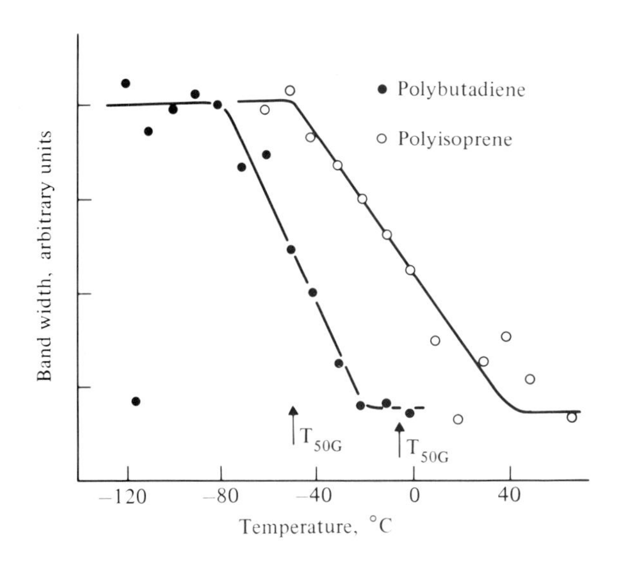

FIGURE 35. Temperature dependence of the ESR line width for polybutadiene (●) and polyisoprene (○).

The electron spin resonance measurements were made at the high frequency of 10^9 Hz and the relaxation, T_g (determining the line width) occurs at temperatures higher than the experimental values, T_{50G}. In order to find a relationship between T_{50G} and T_g, Kumler and Boyer [198] plotted the known T_g values vs. T_{50G} for a number of polymers and copolymers of various structural types and for a wide range of T_g values (Figure 36).

The glass transition temperature of polymers can be determined from the T_{50G} value, on the basis of a curve correlating the two values, T_{50G} and T_g.

From the curve plotted in Figure 36, for $T_{50G} = -4.8°C$, the corresponding glass-transition temperature for synthetic polyisoprene is $T_g = -72°C$, which is virtually identical to the $-70°C$ for Hevea rubber reported in literature [199].

At $T_{50G} = -50°C$ for *cis*-1,4-polybutadiene, the corresponding T_g value determined on the graph given in Figure 36 is $-113°C$, which is also very close to data reported in literature of about $-110°C$.

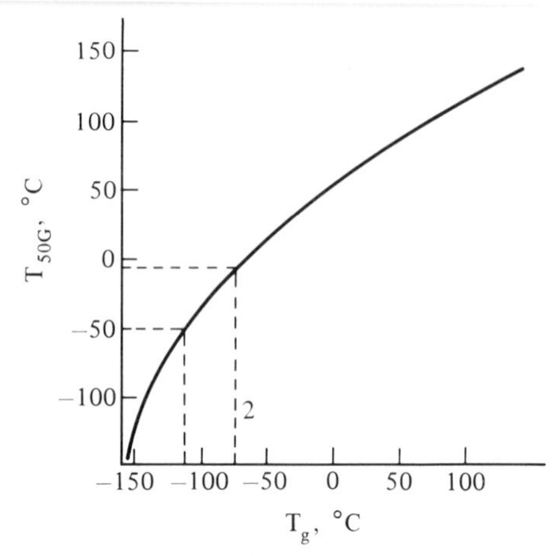

FIGURE 36. Relationship between T_{50G} and T_g, after Kumler and Boyer.

It can be concluded that ESR is a convenient method for the determination of the glass-transition temperature, T_g, of polyisoprene and polybutadiene.

3. Electron microscopy studies of natural rubber and synthetic *cis*-1,4-polyisoprene

The morphological characteristics of natural rubber as compared with those of synthetic polyisoprene were investigated by the method of electron microscopy.

The samples were examined by electron microscopy and electron diffraction with a JEM-120 electron microscope operating at an accelerating voltage of 100 kV. In the best cases, the sample thickness was about 2000–3000 Å, which ensures a good stability of the sample in the electron beam. If the sample is thicker, the polymer can be thermally destroyed, owing to the strong inelastic electron scattering.

The micrographs recorded by scanning electron microscopy of thin, microtomed sections of natural rubber (a and b) and synthetic *cis*-1,4-polyisoprene (c and d) are shown in Figure 37. Both samples have a fibre-like structure and contain inclusions. The similar scale-like fine structures appear at higher magnifications. The above observations allowed us to conclude that the internal structures of the two samples are very close.

The electron diffraction analysis of natural rubber and synthetic *cis*-1,4-polyisoprene indicates an amorphous structure, characteristic for elastomers, as shown in Figures 38a and b. These results are in agreement with the X-ray measurements of the same samples discussed in Part I of the book.

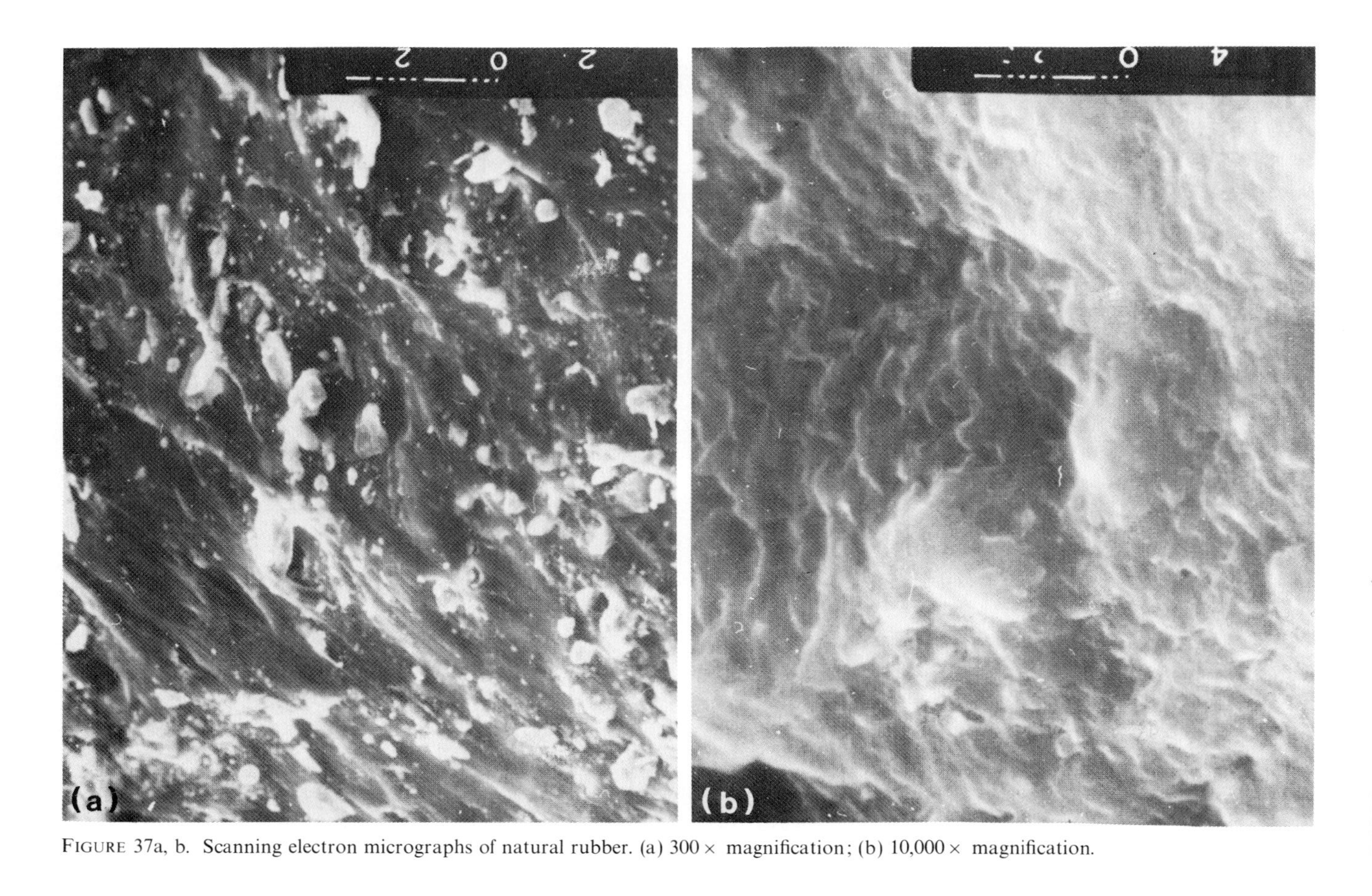

FIGURE 37a, b. Scanning electron micrographs of natural rubber. (a) 300 × magnification; (b) 10,000 × magnification.

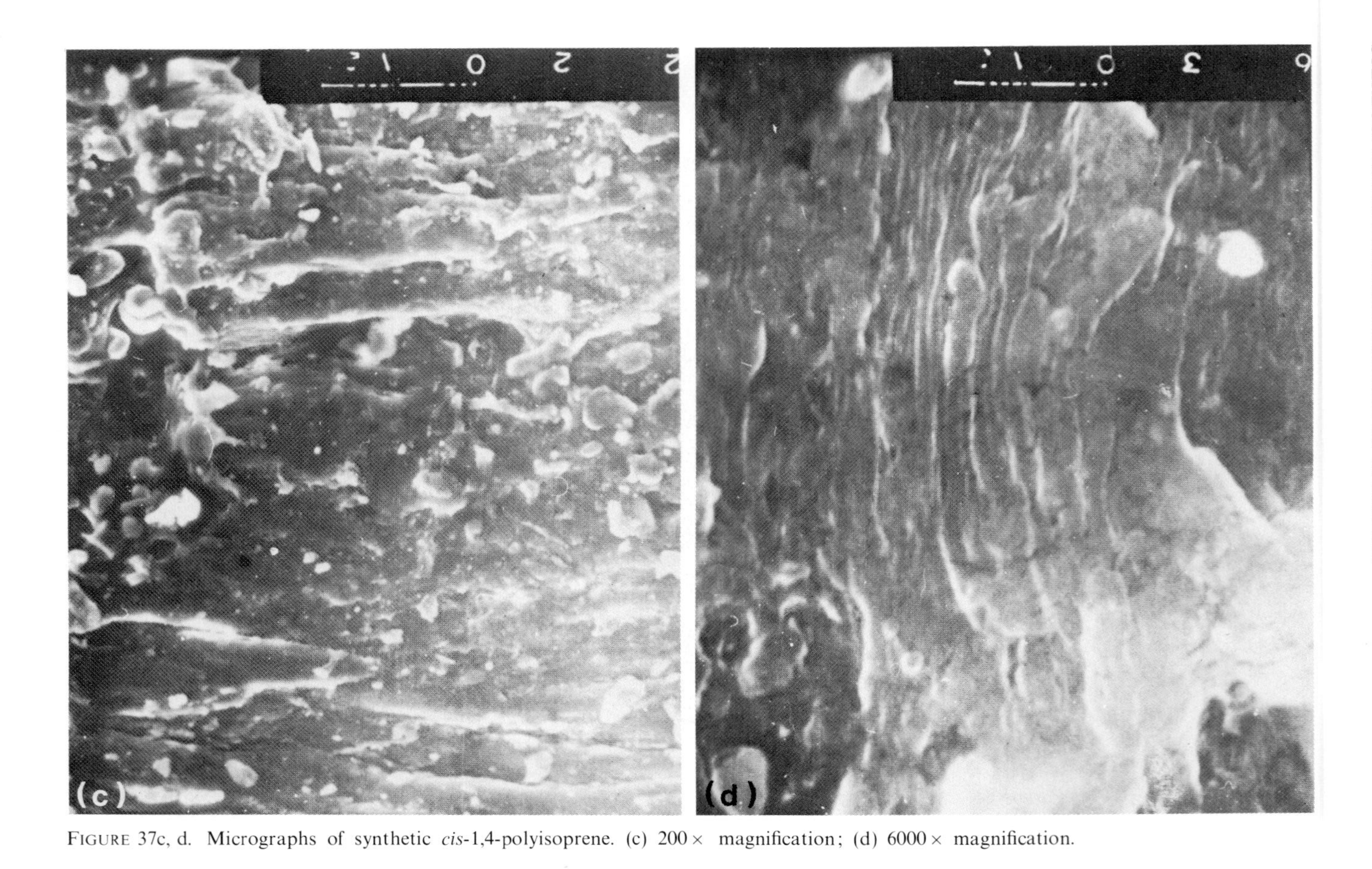

FIGURE 37c, d. Micrographs of synthetic *cis*-1,4-polyisoprene. (c) 200 × magnification; (d) 6000 × magnification.

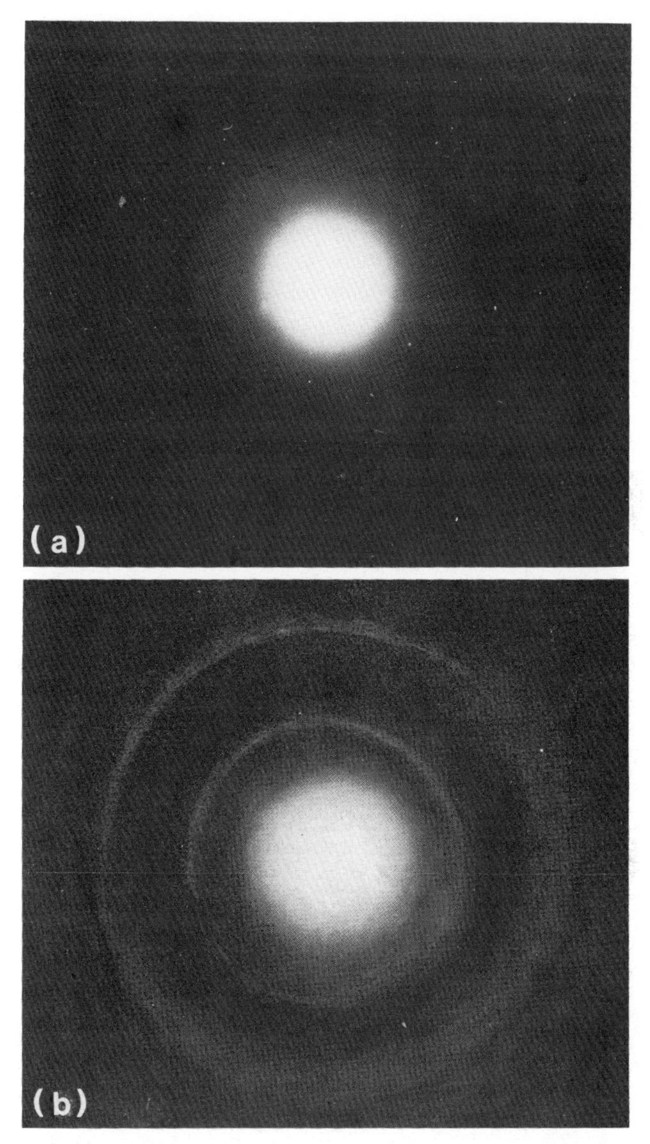

FIGURE 38. Electron diffraction photographs of natural rubber (a) and synthetic *cis*-1,4-polyisoprene (b).

References

Part I

1. S. E. HORNE, *Ind. Eng. Chem.* **48,** 784 (1956).
2. Goodrich-Gulf, Belg. Pat., 543292 (1955).
3. Goodrich-Gulf, Fr. Pat., 1139418 (1957).
4. Goodyear Tire & Rubber Co., *Chem. Eng. News,* **33,** 451 (1955).
5. F. W. STAVELY, *Ind. Eng. Chem.* **48,** 778 (1956).
6. Firestone Tire & Rubber Co., Ital. Pat., 559160 (1956).
7. Firestone Tire & Rubber Co., Ital. Pat., 560200 (1956).
8. G. J. ALLIGER, M. WILLIS and W. A. SMITH, *Rubber World,* **134,** 549 (1956).
9. R. S. ARIES (Phillips Chem. Co.), *Rubber World,* **134,** 719 (1956).
10. G. KRAUS, *Rubber Plast. Age,* **38,** 880 (1957).
11. C. LONGIAVE and R. CASTELLI, *J. Polym. Sci.* **C4,** 387 (1963).
12. K. ZIEGLER, E. HOLZKAMP, H. BREIL and H. MARTIN, *Angew. Chem.* **67,** 541 (1955).
13. H. STAUDINGER, *Die hochmolekularen organischen Verbindungen,* Springer Verlag, Berlin, 1932.
14. C. E. SCHILDKNECHT, *Ind. Eng. Chem.* **39,** 180 (1947).
15. C. E. SCHILDKNECHT and S. T. GROSS, *Ind. Chem. Eng.* **40,** 2104 (1948).
16. C. E. SCHILDKNECHT, A. O. ZOSS and F. GROSSER, *Ind. Eng. Chem.* **41,** 2891 (1949).
17. C. E. SCHILDKNECHT, A. O. ZOSS and F. GROSSER, *Ind. Eng. Chem.* **41,** 1998 (1949).
18. W. COOPER, *Progress in High Polymers,* vol. 1, Ed. J. C. Robb, F. W. Peaker, Heywood Co., London, 1961, p. 281.
19. G. NATTA and R. RIGAMONTI, *Atti reale accad. nazl. Lincei,* **24,** 381 (1936).
20. M. D. SCOTT and J. E. WALKER, *Ind. Eng. Chem.* **32,** 322 (1940).
21. M. L. HUGGINS, *J. Am. Chem. Soc.* **66,** 1991 (1944).
22. C. W. BUNN, *Nature,* **161,** 2929 (1948).
23. C. W. BUNN, *Proc. Roy. Soc.* **A180,** 40 (1947).
24. C. A. KRAUS, U.S. Pat., 2220930 (1940).
25. E. C. HALL and A. W. NASH, *J. Inst. Petrol. Technol.* **23,** 679 (1937).
26. M. FISCHER, Ger. Pat., 874215 (1953).
27. H. W. COOVER, Jr., *J. Polym. Sci.* **C4,** 1511 (1964).
28. K. ZIEGLER, *Angew. Chem.* **76,** 545 (1964).
29. K. ZIEGLER and F. THIELMAN, *Ber. Dtsch. Chem. Ges.* **56,** 1740 (1923).
30. K. ZIEGLER and K. BÄHR, *Ber. Dtsch. Chem. Ges.* **61,** 253 (1928).
31. K. ZIEGLER and H. KLEINER, *Liebigs Ann. Chem.* **473,** 57 (1929).
32. K. ZIEGLER, F. DERSCH and H. WOLLTHAN, *Liebigs Ann. Chem.* **511,** 45 (1934).
33. K. ZIEGLER, *Liebigs Ann. Chem.* **473,** 54 (1929).
34. K. ZIEGLER and L. JAKOB, *Liebigs Ann. Chem.* **511,** 49 (1934).
35. K. ZIEGLER, *Angew. Chem.* **49,** 499 (1936).
36. K. ZIEGLER, A. GRIM and G. R. WILLER, *Liebigs Ann. Chem.* **542,** 90 (1939).
37. K. ZIEGLER, L. JAKOB, H. WOLLTHAN and A. WENZ, *Liebigs Ann. Chem.* **511,** 69 (1934).
38. K. ZIEGLER and H. GELLERT, *Ann.* **567,** 43, 179, 185, 195 (1950).

39. K. Ziegler, *Angew. Chem.* **64,** 323, 326 (1952).
40. K. Ziegler, *Brennstoff Chemie,* **33,** 193 (1952).
41. K. Ziegler, *Brennstoff Chemie,* **35,** 321 (1954).
42. K. Ziegler and H. H. Colonius, *Liebigs Ann. Chem.* **476,** 135 (1930).
43. K. Ziegler and H. G. Gellert, Ger. Pat., 878860 (1953).
44. K. Ziegler and H. G. Gellert, Ger. Pat., 917000 (1954).
45. K. Ziegler, Belg. Pat., 512267 (1952).
46. K. Ziegler, Belg. Pat., 504161 (1951).
47. K. Ziegler and H. G. Gellert, *Petroleum Refiner,* **34,** 8, 111 (1955).
48. K. Ziegler and H. G. Gellert, U.S. Pat., 2695324 (1954).
49. K. Ziegler, Brit. Pat., 742642 (1955).
50. K. Ziegler, Belg. Pat., 527736 (1954).
51. K. Ziegler, Ger. Pat., 1099257 (1955).
52. K. Ziegler, G. Wilke and E. Holzkamp, U.S. Pat., 2781410 (1957).
53. K. Ziegler, Brit. Pat., 773536 (1957).
54. K. Ziegler, Brit. Pat., 777152 (1957).
55. K. Ziegler, Belg. Pat., 540136 (1955).
56. K. Ziegler, E. Holzkamp, A. Breil and H. Martin, *Angew. Chem.* **67,** 426 (1955).
57. G. Natta and P. Corradini, *Angew. Chem.* **68,** 393 (1956).
58. G. Natta, P. Pino and G. Mazzanti, Brit. Pat., 810023 (1959).
59. G. Natta, *Atti accad. nazl. Lincei, Memorie* (8), **4,** 61 (1955).
60. G. Natta, *J. Polym. Sci.* **16,** 143 (1955).
61. G. Natta, P. Pino, P. Corradini, F. Danusso and G. Moraglio, *J. Am. Chem. Soc.* **77,** 1708 (1955).
62. G. Natta, P. Pino and G. Mazzanti, *Chim. Ind.* **37,** 927 (1955).
63. G. Natta, *Experientia,* Suppl., **7,** 21 (1957).
64. G. Natta and P. Corradini, *Rubber Chem. Technol.* **33,** 703 (1960).
65. D. H. De Boer, *Trans. Faraday Soc.* **32,** 10 (1936).
66. G. Natta, P. Corradini and M. Cesari, *Rend. Accad. nazl. Lincei,* **8,** 21, 365 (1956).
67. R. Castelli, *Mat. Plast. Elast.* **29,** 1088 (1963).
68. M. L. Huggins, G. Natta, V. Desreux and H. Mark, *J. Polym. Sci.* **56,** 153 (1962).
69. F. Eirich and H. Mark, *J. Colloid. Sci.* **11,** 748 (1956).
70. I. Stille, *Chem. Reviews,* **58,** 541 (1958).
71. K. Koulen, *Kunststoffe Plastics,* **5,** 149 (1958).
72. B. D. Coleman, *J. Polym. Sci.* **31,** 155 (1958).
73. G. Natta, *Chim. Ind.* **46,** 397 (1964).
74. G. Natta, M. Farina and M. Peraldo, *Rend. Accad. nazl. Lincei,* **8,** 25, 424 (1958).
75. M. Peraldo and M. Farina, *Chim. Ind.* **42,** 1349 (1960).
76. N. G. Gaylord and H. F. Mark, *Linear and Stereoregular Addition Polymers Polymerization with Controlled Propagation,* Polymer Reviews, vol. 2, Interscience Publishers, Inc., New York, 1959.
77. G. B. Butler, *Macromolecular Chemistry,* vol. 2, Butterworth, London, 1962, p. 229.
78. E. J. Badin, *J. Am. Chem. Soc.,* **82,** 86 (1960).
79. T. G. Fox, B. S. Garret, W. E. Goode, S. Gratch, A. Aspell and J. D. Stroupe, *J. Am. Chem. Soc.* **80,** 1768 (1958).
80. N. G. Gaylord and H. F. Mark, *Makromol. Chem.* **44–46,** 448 (1961).
81. E. Perry, *Makromol. Chem.* **65,** 145 (1963).
82. G. Natta, *Chim. Ind.* **38,** 9, 758 (1956).
83. G. Natta, *J. Polym. Sci.* **34,** 3 (1959).
84. G. Hesse, *Katalyse über Komplexe Kationen und Anionen,* 4/2S.61 in *Methoden der organischen Chemie,* Houben-Weil, Stuttgart, 1955.
85. G. Natta, P. Corradini and L. Porri, *Rend. Accad. nazl. Lincei,* **8,** 40, 728 (1956).
86. G. Natta, L. Porri and G. Mazzanti, Belg. Pat., 545942 (1955).
87. G. Natta, *Rend. Accad. nazl. Lincei,* **8,** 28, 442 (1960).
88. H. Hoppf and H. Elias, *Chimia,* **15,** 479 (1961).
89. G. Natta, L. Porri and A. Mazzei, *Chim. Ind.* **39,** 653 (1957).
90. W. Duck, *Rubber & Plastics Weekly,* **143,** 246 (1962).
91. G. Natta, M. Farina, M. Peraldo and M. Donatti, *Chim. Ind.* **42,** 1363 (1960).

92. K. ZIEGLER, Belg. Pat., 549554 (1959).
93. G. NATTA, I. PASQUON and E. GIACHETTI, *Chim. Ind.* **39,** 993 (1957).
94. G. NATTA, *Chim. Ind.* **40,** 267 (1958).
95. G. NATTA, *Chim. Ind.* **39,** 1002 (1957).
96. G. NATTA, *Chim. Ind.* **40,** 97 (1958).
97. G. NATTA, *Chim. Ind.* **40,** 103 (1958).
98. K. C. TSOU, J. E. MEGEE and A. MALATESTA, *J. Polym. Sci.* **58,** 299 (1962).
99. F. DANUSSO, *Chim. Ind.* **44,** 611 (1962).
100. M. H. LEHR and P. H. MOYER, *J. Polym. Sci.* **A3,** 231 (1965).
101. G. NATTA and G. MAZZANTI, *Tetrahedron,* **8,** 86 (1960).
102. D. S. BRESLOW and N. R. NEWBURG, *J. Am. Chem. Soc.* **79,** 5072 (1957).
103. D. S. BRESLOW and N. R. NEWBURG, *J. Am. Chem. Soc.* **81,** 81 (1959).
104. H. UELZMAN, *J. Polym. Sci.* **32,** 457 (1958).
105. H. UELZMAN, *J. Polym. Sci.* **37,** 561 (1959).
106. H. UELZMAN, *J. Org. Chem.* **25,** 671 (1960).
107. H. BESTIAN, C. KLAUS and H. JENSEN, *Angew. Chem., Intern. Ed.* **2,** 32 (1963).
108. C. D. NENITZESCU, C. HUCH and A. HUCH, *Angew Chem.* **68,** 438 (1956).
109. P. COSSEE, *J. Catalysis,* **3,** 80 (1964).
110. E. J. ARLMAN, *J. Catalysis,* **3,** 89 (1964).
111. E. J. ARLMAN and P. COSSEE, *J. Catalysis,* **3,** 99 (1964).
112. E. J. ARLMAN, *J. Polym. Sci.* **62,** 530 (1962).
113. J. BOOR Jr., *J. Polym. Sci.* **C1,** 257 (1963).
114. W. L. CARRICK, *J. Am. Chem. Soc.* **80,** 6455 (1958).
115. W. L. CARRICK and R. W. KLUIBER, *J. Am. Chem. Soc.* **82,** 3863 (1960).
116. W. L. CARRICK, W. T. REICHE and J. J. SMITH, *J. Am. Chem. Soc.* **82,** 3887 (1960).
117. I. OITA and T. D. NEWITT, *J. Polym. Sci.* **43,** 585 (1960).
118. C. H. BEERMAN and H. BESTIAN, *Angew. Chem.* **71,** 618 (1959).
119. A. S. MATLOCK and D. S. BRESLOW, *J. Polym. Sci.* **A3,** 2853 (1965).
120. H. BESTIAN, *Angew. Chem.* **74,** 955 (1962).
121. B. D. BABITSKII and B. A. DOLGOPLOSK, *Dokl. Akad. Nauk SSSR,* **161,** 583 (1965).
122. S. S. MEDVEDEV and A. R. GANTMACHER, *J. Polym. Sci.* **C4,** 173 (1964).
123. V. MARCONI, A. MAZZEI, S. CUCINELLA and M. DE MALDE, *Makromol. Chem.* **71,** 118 (1964).
124. J. BOOR Jr., ACS Div. Polym. Chem., Polymer Preprints, 6/2, 890 (1965).
125. G. NATTA, I. PASQUON, A. ZAMBELLI and G. GATTJ, *Makromol. Chem.* **70,** 206 (1964).
126. J. BOOR Jr. and E. A. YOUNGMAN, *J. Polym. Sci.* **B2,** 265 (1964).
127. W. COOPER, D. E. EAVES, D. G. OWEN and G. VAUGHAN, *J. Polym. Sci.* **C4,** 211 (1964).
128. G. A. RAZUVAEV and K. S. MINSKER, *Vysokomol. Soed.* **1,** 1969 (1959).
129. G. A. RAZUVAEV and K. S. MINSKER, *J. Polym. Sci.* **52,** 299 (1961).
130. G. A. RAZUVAEV and K. S. MINSKER, *J. Polym. Sci.* **44,** 285 (1960).
131. K. VESELY, J. AMBROZ and R. VILIM, *J. Polym. Sci.* **55,** 25 (1961).
132. F. DAWANS and P. TEYSSIÉ, *Bull. Soc. Chim. France,* **10,** 2376 (1963).
133. L. M. LANOVSKAYA, N. V. MAKLETSOVA, A. R. GANTMAKHER and S. S. MEDVEDEV, *Vysokomol. Soed.* **7,** 741 (1965).
134. G. NATTA, I. PASQUON and A. ZAMBELLI, *Makromol. Chem.* **70,** 191 (1964).
135. G. NATTA, *Khimia i tekhnologiya polimerov,* **7,** 8, 112 (1960).
136. F. DANUSSO, *J. Polym. Sci.* **C4,** 1497 (1964).
137. G. NATTA, *J. Polym. Sci.* **34,** 21 (1959).
138. W. COOPER, *Progress in High Polymers,* Ed. J. C. Robb, Heywood Co., London, 1961, p. 281.
139. G. NATTA and I. PASQUON, *Advances in Catalysis,* vol. XI, Acad. Press, New York, 1959, p. 1.
140. G. NATTA, P. CORRADINI and C. ALLEGRA, *J. Polym. Sci.* **51,** 399 (1961).
141. M. I. MOSEVITSKII, *Uspekhi khimii,* **28,** 465 (1959).
142. P. COSSEE, *Trans. Faraday Soc.* **56,** 1226 (1962).
143. J. FURUKAWA, *Bull. Inst. Chem., Res. Kyoto Univ.* **40,** 130 (1962).
144. A. V. TOPCHIEV, B. A. KRENTSEL' and L. G. SIDOROVA, *Uspekhi khimii,* **30,** 192 (1961).
145. A. V. TOPCHIEV and B. A. KRENTSEL', *Dokl. Akad. Nauk SSSR,* **128,** 193 (1959).
146. A. V. TOPCHIEV, B. A. KRENTSEL' and L. G. SIDOROVA, *Izv. Akad. Nauk. SSSR, ser. Khim.* **22,** 1133 (1958).
147. C. L. ARCUS, *Progress in Stereochemistry,* vol. 3, 1960, p. 264.

148. J. L. Mateo, *Rev. Plast. Mat.* **103,** 1 (1965).
149. W. M. Saltman, *J. Polym. Sci.* **A1,** 373 (1963).
150. A. Clark, J. P. Hogan, R. L. Banks and W. C. Laning, *Ind. Eng. Chem.* **48,** 1162 (1956).
151. M. R. Cines, G. H. Dale and E. W. Mellow, *Chem. Eng. Progress,* **54,** 95 (1958).
152. Ger. Pat., 1051004 (1954).
153. A. Zletz, Standard Oil (Indiana), U.S. Pat., 2692257 (1954).
154. Standard Oil (Indiana), U.S. Pat., 2802814 (1957).
155. F. F. Peters, A. Zletz and B. L. Evening, *Ind. Eng. Chem.* **49,** 1879 (1957).
156. E. Field and M. Feller, *Ind. Eng. Chem.* **49,** 1883 (1957).
157. E. Field and M. Feller, Standard Oil (Indiana), Ger. Pat., 1057339 (1953).
158. E. Field and M. Feller, Standard Oil (Indiana), Ger. Pat., 1040245 (1953).
159. E. Field and M. Feller, Standard Oil (Indiana), Ger. Pat., 1027878 (1953).
160. E. Field and M. Feller, Standard Oil (Indiana), Ger. Pat., 1031581 (1953).
161. E. F. Peters and B. L. Evening, Standard Oil (Indiana), Ger. Pat., 1069386 (1956).
162. E. F. Peters and B. L. Evening, Standard Oil (Indiana), Fr. Pat., 1142344 (1956).
163. A. S. Seelig, Standard Oil (Indiana), Ger. Pat., 1026530 (1953).
164. A. A. Morton and F. D. Marsch, *J. Am. Chem. Soc.* **71,** 487 (1949).
165. A. A. Morton and F. D. Marsch, *J. Am. Chem. Soc.* **72,** 3785 (1950).
166. A. A. Morton and E. Grovenstein, *J. Am. Chem. Soc.* **74,** 5437 (1952).
167. A. A. Morton, *Ind. Eng. Chem.* **42,** 1488 (1950).
168. D. Braun and W. Kern, *J. Polym. Sci.* **C4,** 197 (1964).
169. D. Braun, M. Herner and W. Kern, *Adv. Chem.* **24,** 120 (1962).
170. M. Szwarc, *Chem. Ind. (Rev.)*, 1589 (1958).
171. T. Higashimura, T. Watanabe, K. Zugucki and S. Okamura, *J. Polym. Sci.* **C4,** 361 (1964).
172. C. E. H. Bawn, A. M. North and J. S. Walker, *Polymer,* **5,** 8, 419 (1964).
173. W. Cooper, *Rubber Plast. Age,* **44,** 44 (1963).
174. H. Sinn, K. Ludborg and C. Kirchen, *Angew. Chem.* **70,** 744 (1958).
175. R. S. Stearns and L. E. Forman, *J. Polym. Sci.* **41,** 381 (1959).
176. C. E. H. Bawn, *Rubber Plast. Age,* **42,** 267 (1961).
177. M. Roha, *Fortsch. Hochpolym. Forsch.* **1,** 512 (1960).
178. C. E. H. Bawn and A. Ledwith, *Quarterly Revs.* **16,** 361 (1962).
179. K. F. O'Driscoll and A. V. Tobolsky, *J. Polym. Sci.* **35,** 259 (1959).
180. T. L. Brown, R. L. Gerteis, D. A. Bafus and J. A. Ladd, *J. Am. Chem. Soc.* **86,** 2135 (1964).
181. D. Margerison and J. P. Newport, *Trans. Faraday Soc.* **59,** 2058 (1963).
182. T. L. Brown, *J. Organometal. Chem.* **5,** 2, 191 (1966).
183. H. Sinn and C. Lundborg, *Makromol. Chem.* **70,** 251 (1964).
184. R. Wack and P. West, *J. Organometal. Chem.* **2,** 2, 198 (1966).
185. S. Bywater and D. J. Worsfold, *Can. J. Chem.* **40,** 1564 (1962).
186. M. Morton and D. J. Fetters, *J. Polym. Sci.* **A2,** 311 (1964).
187. A. G. Evans and D. B. George, *J. Chem. Soc., Oct.* 4653 (1961).
188. R. Cubbon and C. P. D. Margerison, *Proc. Roy. Soc.* **A268,** 260 (1962).
189. H. Sinn and F. Patat, *Angew. Chem., Intern. Ed.* **3,** 2 (1964).
190. J. Bercowitz, D. A. Bafus and T. L. Brown, *J. Phys. Chem.* **65,** 1380 (1961).
191. M. Weiner, C. Vogel and R. West, *Inorg. Chem.* **1,** 654 (1962).
192. H. L. Hsieh, *J. Polym. Sci.* **A3,** 153 (1965).
193. H. Sinn and G. T. Onsager, *Makromol. Chem.* **55,** 167 (1962).
194. D. S. Breslow and N. R. Newburg, *Chem. Eng. News,* **36,** 29, 56 (1958).
195. J. Minoux, *Makromol. Chem.* **44–46,** 519 (1961).
196. J. P. Kennedy, *Fortsch. Hochpolym. Forsch.* **3,** 508 (1964).
197. Rohm & Hass Co., Belg. Pat., 566713 (1958).
198. H. Wexler and J. Manson, *J. Polym. Sci.* **A3,** 2965 (1965).
199. C. Longiave and L. Porri, *Mat. Plast. Elast.* **29,** 695 (1963).
200. J. G. Balas and L. M. Porter, U.S. Pat., 3040016 (1962).
201. Montecatini, Brit. Pat., 916643 (1963).
202. Montecatini, Brit. Pat., 924244 (1963).
203. Montecatini, Brit. Pat., 924427 (1963).
204. Montecatini, Brit. Pat., 936061 (1963).

205. Montecatini, Ital. Pat., 592477 (1957).
206. Montecatini, Ital. Pat., 594618 (1958).
207. J. G. BALAS, Shell Oil Co., U.S. Pat., 3067189 (1962).
208. L. M. PORTER and J. G. BALAS, U.S. Pat., 30662216 (1962).
209. H. TUCKER, U.S. Pat., 3094514 (1963).
210. E. A. YOUNGMAN, U.S. Pat., 3066128 (1962).
211. E. A. YOUNGMAN and J. BOOR, U.S. Pat., 3084148 (1963).
212. G. NATTA, L. PORRI and O. CARBONARO, *Atti Accad. nazl. Lincei R. C. Cl. Sci. Fiz. Mat.* **29,** 491 (1960).
213. F. P. VAN DE KAMP, *Makromol. Chem.* **93,** 202 (1966).
214. C. E. H. BAWN, *Rubber Plast. Age*, **46,** 510 (1965).
215. H. SCOTT, R. F. BELT and O. E. O'REILLI, *J. Polym. Sci.* **A2,** 2243 (1964).
216. L. PORRI, G. NATTA and G. M. GALLAZZI, *Chim. Ind.* **46,** 429 (1964).
217. B. D. BABITSKII and B. A. DOLGOPLOSK, *Dokl. Akad. Nauk SSSR*, **161,** 583 (1965).
218. E. O. FISCHER and G. G. BÜRGER, *Z. Naturforschung*, **16b,** 77 (1961).
219. G. VILKE, *Uspekhi Khimii*, **33,** 687 (1964).
220. B. D. BABITSKII and B. A. DOLGOPLOSK, *Vysokomol. Soed.* **6,** 12 (1964).
221. B. D. BABITSKII and B. A. DOLGOPLOSK, *Dokl. Akad. Nauk. SSSR*, **160,** 4 (1965).
222. M. FARINA and M. RAGGAZZINI, *Chim. Ind.* **40,** 816 (1958).
223. C. E. N. BAWN and R. SYMCOX, *J. Polym. Sci.* **34,** 139 (1959).
224. J. S. LASKY, H. K. GARNER and R. H. EDWART, *Ind. Eng. Chem. Prod. Res. Develop.* **1,** 83 (1962).
225. D. H. DAWES and C. A. WINKLER, *J. Polym. Sci.* **A2,** 3029 (1964).
226. G. NATTA, G. DALL'ASTA and G. MOTRONI, *J. Polym. Sci.* **B2,** 349 (1964).
227. U.S. Rubber Co., U.S. Pat., 3025286 (1962).
228. R. E. RINEHART, H. P. SMITH, H. S. WITT and H. ROMEYN, *J. Am. Chem. Soc.* **83,** 4863 (1961).
229. Shell Intern. Rev. Matsch., Belg. Pat., 604903 (1961).
230. A. J. CANALE, W. A. HEWETT, T. M. SHRINE and E. A. YOUNGMAN, *Chem. Ind.* **24,** 1054 (1962).
231. G. NATTA, G. DALL'ASTA and L. PORRI, *Makromol. Chem.* **81,** 253 (1965).
232. J. CHATT and L. M. VENANZI, *J. Am. Chem. Soc.* **79,** 4735 (1957).
233. P. TEYSSIÉ and H. DAUBY, *J. Polym. Sci.* **B2,** 413 (1964).
234. M. MORTON, I. PIIRMA and B. DAS, *Rev. Gen. Caout.* **42,** 267 (1965).
235. D. B. LUDLUM, A. W. ANDERSON and C. E. ASHBY, *J. Am. Chem. Soc.* **80,** 1380 (1958).
236. J. BADIN, *J. Am. Chem. Soc.* **80,** 6545 (1958).
237. F. EIRICH and H. MARK, *Kunststoffe Plastics*, **3,** 136 (1956).
238. J. FURUKAWA and T. TSURUTA, *J. Polym. Sci.* **36,** 275 (1959).
239. P. SIGWALT, *Plast. Mod. Elast.* **17,** 7, 117 (1965).
240. T. G. FOX, W. E. GOODE, S. GRATCH, C. M. HUGHETT, A. SPELL and J. D. STROUPE, *J. Polym. Sci.* **31,** 173 (1958).
241. J. P. MAYNARD and W. E. MACHEL, *J. Polym. Sci.* **13,** 235 (1954).
242. W. KERN, *Chem. Zeit.* **87,** 22, 799 (1963).
243. F. DANUSSO, *Makromol. Chem.* **35A,** 116 (1960).
244. A. G. EVANS, *Nature*, **157,** 102 (1946).
245. K. FUKUI and T. HIGASHIMURA, *J. Polym. Sci.*, **39,** 487 (1959).
246. M. SZWARC, M. LEVY and R. MOLCOVICH, *J. Am. Chem. Soc.* **78,** 2656 (1956).
247. M. SZWARC, *Makromol. Chem.* **35,** 132 (1960).
248. A. V. TOBOLSKY, D. J. KELLEY, K. F. O'DRISCOLL and C. E. ROGERS, *J. Polym. Sci.* **28,** 425 (1958).
249. A. GILCHRIST, *J. Polym. Sci.* **34,** 62 (1959).
250. J. L. R. WILLIAM, T. M. LAAKSO and E. J. DULMAGE, *J. Org. Chem.* **23,** 638 (1958).
251. W. KERN, D. BRAUN and M. HERNER, *Makromol. Chem.* **28,** 66 (1958).
252. D. BRAUN and W. KERN, *Gummi Asbest Kunst.* **16,** 968 (1963).
253. J. L. R. WILLIAMS, J. VAN DEN BERGHE, W. J. DULMAGE and K. R. DURHAM, *J. Am. Chem. Soc.* **78,** 1260 (1956).
254. J. L. R. WILLIAMS, J. VAN DEN BERGHE, W. J. DULMAGE and K. R. DURHAM, *J. Am. Chem. Soc.* **79,** 1716 (1957).
255. A. A. MORTON and L. D. TAYLOR, *J. Polym. Sci.* **38,** 7 (1959).

256. A. A. Morton and F. K. Ward, *J. Org. Chem.* **24,** 929 (1959).
257. D. Braun, H. Herner and W. Kern, *Makromol. Chem.* **36,** 232 (1960).
258. C. L. Lee, J. Smith and M. Szwarc, *Trans. Faraday Soc.* **59,** 1192 (1963).
259. D. Braun and W. Fischer, *Makromol. Chem.* **85,** 155 (1965).
260. A. A. Korotkov and Zh. A. Silaev, *Vysokomol. Soed.* **1,** 443 (1959).
261. A. A. Korotkov and Zh. A. Silaev, *J. Polym. Sci.* **39,** 565 (1959).
262. S. Schlik and M. Levy, *J. Phys. Chem.* **64,** 883 (1960).
263. K. F. O'Driscoll and A. V. Tobolsky, *J. Polym. Sci.* **31,** 115 (1958).
264. Yu. L. Spirin, *J. Polym. Sci.* **58,** 1181 (1962).
265. N. Chernova and G. Mikhailov, *Dokl. Akad. Nauk SSSR,* **111,** 2249 (1956).
266. F. J. Welch, *J. Am. Chem. Soc.* **82,** 600 (1960).
267. G. E. Coates, *Organometallic Compounds*, London, 1956, p. 1.
268. E. Warhurst, *Discussion Faraday Soc.* **2,** 239 (1947).
269. G. Wittig and E. Stahnecker, *Liebig Ann. Chem.* **605,** 69 (1957).
270. D. J. Worsfold and S. Bywater, *J. Chem. Soc.* 523 (1960).
271. K. Buttler, *J. Polym. Sci.* **48,** 357 (1960).
272. I. Kuntz, *J. Polym. Sci.* **42,** 299 (1960).
273. D'Alelio and T. Miranda, *Chem. Ind.* **5,** 163 (1959).
274. E. Perry, *Makromol. Chem.* **65,** 145 (1963).
275. H. N. Friedländer and K. Oita, *Ind. Eng. Chem.* **49,** 1885 (1957).
276. E. W. Duck, *J. Polym. Sci.* **34,** 86 (1959).
277. G. Natta, *Khimiia i tekhnologiia polimerov,* **7,** 8, 112 (1960).
278. J. Imanischi and S. Okamura, *Makromol. Chem.* **48,** 246 (1961).
279. C. W. Childers, *J. Am. Chem. Soc.* **85,** 229 (1963).
280. H. Sinn, H. Winter and W. V. Tirpitz, *Makromol. Chem.* **48,** 59 (1961).
281. G. Natta, *Angew. Chem.* **67,** 430 (1955).
282. G. Natta, *Mod. Plast.* **34,** 169 (1956).
283. M. Szwarc, *Makromol. Chem.* **35A,** 116 (1960).
284. G. Natta, *Makromol. Chem.* **16,** 213, (1955).
285. G. Natta, *Chim. Ind.* **38,** 751 (1956).
286. E. Patat and H. Sinn, *Angew. Chem.* **70,** 496 (1958).
287. P. Cossee, *Tetrahedron Letters,* **17,** 12 (1960).
288. P. H. De Bruyn, *Chem. Week,* **56,** 161 (1960).
289. J. W. C. Chien, *J. Am. Chem. Soc.* **80,** 6455 (1958).
290. C. van Heerderen, *J. Polym. Sci.,* **34,** 46 (1956).
291. G. Natta, P. Pino, G. Mazzanti, U. Giannini, E. Mantica and M. Pelardo, *J. Polym. Sci.* **26,** 120 (1957).
292. H. Jones, U. Martins and M. P. Thorne, *Can. J. Chem.* **38,** 2303 (1960).
293. D. McGowan and B. M. Ford, *J. Chem. Soc.* **3,** 1149 (1958).
294. N. L. Carrick, A. G. Chasar and J. J. Smith, *J. Am. Chem. Soc.* **82,** 5319 (1960).
295. M. Gippin, *Ind. Eng. Chem. Prog. Res. Develop.* **4,** 3, 160 (1965).
296. J. G. Balas and De la Mare, *J. Polym. Sci.* **A3,** 2243 (1965).
297. G. Natta and L. Porri, *Makromol. Chem.* **71,** 207 (1964).
298. W. Cooper, E. Eaves and G. Vaughan, *Makromol. Chem.* **67,** 229 (1963).
299. P. H. Plesch, *The Chemistry of Cationic Polymerization*, Pergamon Press Ltd., London, 1963.
300. C. M. Fontana, *The Chemistry of Cationic Polymerization*, Ed. P. H. Plesch, Pergamon Press, London, 1963, Chap. 5, p. 211.
301. M. N. Berger and B. M. Grieveson, *Makromol. Chem.* **83,** 80 (1965).
302. P. E. M. Allen and J. F. Harrod, *Makromol. Chem.* **32,** 153 (1959).
303. J. Minoux and J. Parrod, *Compt. Rend.* **252,** 19, 2887 (1961).
304. M. Morton, E. E. Bostick and R. Rivigni, *Rubber Plast. Age,* **42,** 397 (1961).
305. Hung-Chuan Hsieh, Ping-Hsin Chin and Hsin-Chung Tan, *Ko Fen Tzu T'ung Hsun,* **6,** 312 (1964); *C.A.* **63,** 13418h.
306. J. C. W. Chien, *J. Am. Chem. Soc.* **81,** 86 (1959).
307. J. Lal, *J. Polym. Sci.* **31,** 179 (1958).
308. K. Kukui, *J. Polym. Sci.* **37,** 341, 353 (1959).

309. K. VESELY, J. AMBROZ and O. HAMRIK, *J. Polym. Sci.* **C4,** 11 (1964).
310. I. PASQUON and M. NARDUSSI, *Chim. Ind.* **41,** 387, 534 (1959).
311. G. NATTA, I. PASQUON and E. GIACHETTI, *Makromol. Chem.* **24,** 258 (1957).
312. T. TOKUJI and T. KEII, *Kogyo Kagaku Zasshi,* **67,** 1433 (1964); *C.A.* **63,** 5745a.
313. E. J. BADIN, *J. Am. Chem. Soc.* **80,** 6549 (1958).
314. S. M. BURNETT and J. P. T. TAIT, *J. Polym. Sci.* **34,** 46 (1959).
315. F. DANUSSO and D. SIANESI, *Chim. Ind.* **40,** 909 (1958).
316. R. L. CLELAND, R. T. LETSINGER and E. E. MAGAT, *J. Polym. Sci.* **39,** 249 (1959).
317. W. M. SALTMAN, W. E. GIBBS and J. LAL, *J. Am. Chem. Soc.* **80,** 5615 (1958).
318. W. A. MARCONI, A. MAZZEI, M. ARALDI and M. DE MALDE, *J. Polym. Sci.* **A3,** 153 (1965).
319. HONG-CHUA HSIEH, SHIN-CHUNG SHEN and YING-TAI CHIN, *Sci. Sinica (Peking),* **14,** 485 (1965); *C.A.* **62,** 16387e.
320. HUNG-CHUA HSIEH and HSIN-CHUNG TAN, *Ko Fen Tzu T'ung Hsun,* **7,** 18 (1965); *C.A.* **63,** 16579h.
321. L. H. TUNG, *J. Polym. Sci.* **20,** 495 (1956).
322. H. N. WESSLAU, *Makromol. Chem.* **20,** 111 (1956).
323. R. S. ARIES and A. P. SACHS, *J. Polym. Sci.* **21,** 551 (1959).
324. P. S. FRANCIS, R. C. COOKE and J. H. ELLIOT, *J. Polym. Sci.* **31,** 453 (1958).
325. L. H. TUNG, *J. Polym. Sci.* **24,** 333 (1957).
326. B. M. GRIEVESON, *Makromol. Chem.* **84,** 93 (1965).
327. S. E. BRESLER and M. I. MOSEVITSKII, *Dokl. Akad. Nauk. SSSR,* **121,** 859 (1958).
328. H. E. DIEM, H. TUCKER and C. F. GIBBS, Abstract, 133, Meeting, *Amer. Chem. Soc.,* New York, Sept. 1957, p. 9u.
329. N. G. GAYLORD, T. K. KROLL and H. MARK, *J. Polym. Sci.* **42,** 417 (1960).
330. P. PINO, L. LARDICCI and I. CENTONI, *J. Org. Chem.* **24,** 1399 (1959).
331. H. E. ADAMS, R. S. STEARNS, W. A. SMITH and J. C. BINDER, *Ind. Eng. Chem.* **50,** 1507 (1958).
332. A. M. GUYOR ROCHINA and J. TRAMBOUSE, *Bull. Soc. Chim. France,* **10,** 1893 (1962).
333. M. GIPPIN, *Ind. Eng. Chem. Prod. Res. Develop.* **1,** 32 (1962).
334. A. V. TOBOLSKY and C. E. ROGERS, *J. Polym. Sci.* **40,** 73 (1959).
335. H. SINN, *Angew. Chem.* **75,** 805 (1963).
336. B. A. DOLGOPLOSK, *Dokl. Akad. Nauk SSSR,* **146,** 362 (1962).
337. P. E. M. ALLEN, D. GILL and C. R. PATRICK, *J. Polym. Sci.* **C4,** 127 (1964).
338. G. BIER, *Makromol. Chem.* **70,** 44 (1964).
339. G. LEBMAN and A. GUMBOLDT, *Makromol. Chem.* **70,** 23 (1964).
340. T. P. WILSON and G. F. HURLEY, *J. Polym. Sci.* **C1,** 281 (1963).
341. F. KUKUI, T. KAGIYA and S. MACHI, *Bull. Chem. Soc. Japan,* **35,** 306 (1962).
342. H. FELLCHENFOLD and M. JELESON, *J. Phys. Chem.* **63,** 720 (1959).
343. G. NATTA, *Chim. Ind.* **42,** 1207 (1960).
344. G. NATTA, *Mat. Plast.* **1,** 3 (1958).
345. J. AMBROZ and K. VESELY, *Proc. IUPAC Symposium on Macromolecules Wiesbaden,* 1959, Paper IV A1.
346. G. NATTA, I. PASQUON and G. GIACHETTI, *Angew. Chem.* **69,** 213 (1957).
347. E. H. KOHN, D. L. SHURMANS, J. V. CAVENDER and R. A. MENDELSON, *J. Polym. Sci.* **58,** 681 (1962).
348. G. NATTA, L. PORRI and A. MAZZEI, *Chim. Ind.* **41,** 398 (1959).
349. V. N. ZONIK and B. A. DOLGOPLOSK, *Vysokomol. Soed.* **7,** 308 (1965).
350. H. L. HSIEH, *J. Polym. Sci.* **A3,** 163 (1965).
351. A. F. JOHNSON, D. J. WORSFOLD and S. BYWATER, *J. Polym. Sci.* **A3,** 449 (1965).
352. W. M. SALTMAN, *J. Polym. Sci.* **46,** 375 (1960).
353. A. A. KOROTKOV, *Vysokomol. Soed.* **1,** 46 (1959).
354. L. J. FETTERS, *J. Res. Natl. Bur. Std.* **69A,** 159 (1965).
355. H. L. HSIEH, *J. Polym. Sci.* **A3,** 153 (1965).
356. B. FRANÇOIS, H. SINN and J. PARROD, *J. Polym. Sci.* **C4,** 375 (1964).
357. K. F. O'DRISCOLL, E. N. RICHEZZA and J. E. CLARK, *J. Polym. Sci.* **A3,** 3241 (1965).
358. H. SINN and C. LUNDBORG, *Makromol. Chem.* **47,** 87 (1961).
359. G. FRIEDMANN, P. SCHUE, M. BRINI, A. DELUZARCHE and A. MAILAND, *Bull. Soc. Chim. France,* **6,** 1728 (1965).

360. D. S. Breslow, W. P. Long and N. R. Newburg, *Rubber Age (London)*, **41,** 155 (1960).
361. A. I. Medalia, A. Orzechowski, J. A. Tuchera and J. P. Morley, *J. Polym. Sci.* **41,** 249 (1959).
362. F. Danusso, B. Calcaguo and D. Sianesi, *Chim. Ind.* **41,** 287 (1959).
363. A. Schindler, *J. Polym. Sci.* **C4,** 81 (1964).
364. G. Natta, F. Danusso and D. Sianesi, *Makromol. Chem.* **30,** 238 (1959).
365. G. Natta, I. Pasquon and E. Giachetti, *Angew. Chem.* **69,** 213 (1957).
366. J. C. W. Chien, *J. Polym. Sci.* **A1,** 425 (1963).
367. R. C. P. Cubbon and D. Margerison, *Proc. Chem. Soc.* 146 (1960).
368. D. J. Worsfold and S. Bywater, *Can. J. Chem.* **38,** 1891 (1960).
369. J. Welch, *J. Am. Chem. Soc.* **81,** 1345 (1959).
370. E. N. Kropacheva, B. A. Dolgoplosk and E. M. Kuznetsova, *Dokl. Akad. Nauk SSSR*, **130,** 1253 (1960).
371. M. Morton, J. L. Fetters and E. Bostick, *J. Polym. Sci.* **C1,** 311 (1963).
372. C. Lundborg and H. Sinn, *Makromol. Chem.* **41,** 242 (1960).
373. M. Morton, *J. Polym. Sci.* **A2,** 311 (1964).
374. H. Sinn and G. Z. Onsager, *Makromol. Chem.* **52,** 167 (1962).
375. M. Morton, *J. Polym. Sci.* **A1,** 1735 (1963).
376. H. Sinn and C. Lundborg, *Makromol. Chem.* **47,** 242 (1961).
377. F. Schune and G. Friedmann, *Bull. Soc. Chim. France*, **12,** 632 (1965).
378. S. Bywater and D. Worsfold, *Can. J. Chem.* **40,** 1564 (1962).
379. F. Dawans and G. Lefebre, *Rev. Inst. Franc. Petrol*, **11,** 151 (1961).
380. D. Craig, J. Schipman and R. Fouler, *J. Am. Chem. Soc.* **83,** 2885 (1961).
381. R. Kirk and D. F. Othmer, *Encyclopedia of Chemical Technology*, vol. 7, Interscience, New York, 1951, p. 625.
382. K. Uhrig, *Ind. Eng. Chem., Anal. Ed.* **18,** 550 (1946).
383. P. Hersch, *Chim. Anal.* **41,** 189 (1959).
384. E. Berl, *Chemisch-technische Untersuchungsmethoden*, Bd. 5, Springer Berlin, 1932, S. 923.
385. J. Michell Jr. and D. M. Smith, *Chemical Analysis*, vol. 5, Interscience, New York, 1948, *Aquametry*, p. 168.
386. V. Herout, B. Keil, M. Protiva, M. Hudlicky, I. Ernest and J. Gut, *Technica lucrărilor de laborator în chimia organică*, transl. from Czech., Ed. "Technica", Bucureşti, 1959, p. 416.
387. M. R. Papin, *Bull. Assoc. Franc. Techn. Petrol.* **164,** 141 (1964).
388. Ch. Hersch, *Macromoleculare Sieves*, Reinhold Publ. Corp., New York, 1961, p. 83.
389. F. D. Snell and C. T. Snell, *Colorimetric Methods of Analysis*, vol. 3, D. Van Nostrand Co., New York, 1961.
390. Houben-Weil, *Methoden der Organischen Chemie*, Bd. II, Anal. Meth., G. Tieme Verlag, Stuttgart, 1953, S. 313.
391. W. M. Saltman, Fr. Pat., 1236241 (1960).
392. B. Wargotz, Fr. Pat., 1239727 (1960).
393. R. P. Zelinski and D. R. Smith, Fr. Pat., 1247307 (1960).
394. L. S. Kofman, *Vestn. tekhn. ekonom.* **4,** 10 (1962).
395. K. Ziegler and H. S. Gellert, *Angew. Chem.* **67,** 425 (1955).
396. Bochumer Chemie Imhausen Co., M.B.H., Hamburg, Prospect.
397. E. P. Tepenitsyna, M. I. Farberov and G. I. Levinskaya, *Vysokomol. Soed.* **1,** 1148 (1959).
398. F. J. Welcher, *The Analytical Uses of Ethylendiaminotetraacetic Acid*, D. Van Nostrand Co., Toronto, 1958, p. 168.
399. K. Ziegler and H. S. Gellert, *Ann. Chim.* **20,** 629 (1920).
400. *Ullmans Encyklopädie der Technischen Chemie*, Bd. 17, Urban & Schwarzenberg, München-Berlin-Wien, 1966, p. 420.
401. *Ullmans Encyklopädie der Technischen Chemie*, Bd. 17, Urban & Schwarzenberg, München-Berlin-Wien, 1966, p. 430.
402. R. Orzechowski, *J. Polym. Sci.* **34,** 65 (1959).
403. A. Gilchrist, *J. Polym. Sci.* **34,** 49 (1959).
404. J. Furukawa, T. Tsuruta, T. Fueno, R. Sakata and K. Ito, *Makromol. Chem.* **30,** 213 (1959).
405. P. E. N. Allen and F. J. Harrod, *Makromol. Chem.* **32,** 153 (1959).

406. K. Fukui, T. Kagya, T. Shimidzu, T. Yagi, S. Machi and S. Yuasa, *J. Polym. Sci.* **37,** 341 (1959).
407. A. N. Zelikman, S. V. Samsonov and O. E. Krein, *Metallurgiia redkikh metallov*, Moskva, 1954, p. 197.
408. *Manualul Chimistului*, red. C. Lakner, vol. 1, Ed. "Technica", Bucureşti, 1949, p. 491.
409. F. R. Meyer and G. Rouge, *Angew. Chem.* **52,** 673 (1939).
410. M. Schütze, *Angew. Chem.* **70,** 697 (1958).
411. E. Ceauşescu, S. Bittman and I. Florescu, Rom. Pat., 48237 (1964).
412. N. Yamazaki, T. Suminoe and S. Kambara, *Makromol. Chem.* **65,** 157 (1963).
413. W. M. Saltman, S. F. Farson and E. Schoenberg, *Rubber Plast. Age*, **46,** 502 (1965).
414. J. B. Field, D. E. Woodford and S. D. Gehman, *J. Appl. Phys.* **35,** 386 (1964).
415. W. S. Richardson and A. Sacher, *J. Polym. Sci.* **10,** 353 (1953).
416. J. L. Binder and H. C. Ransaw, *Anal. Chem.* **29A,** 503 (1957).
417. K. V. Nel'son, *Izv. Akad. Nauk SSSR, Ser. Phys.* **6,** 741 (1954).
418. E. I. Pokrovskii and N. V. Volkenstein, *Dokl. Akad. Nauk. SSSR*, **95,** 301 (1954).
419. H. L. Hsieh, *Dissert. Abstr.* **19,** m, 8 (1959).
420. A. R. Kemp and G. S. Mueller, *Ind. Eng. Chem., Anal. Ed.* **6,** 52 (1934).
421. A. V. Tobolsky, *Properties and Structure of Polymers*, Interscience Publ., Inc., New York, 1960.
422. H. P. Klug and L. E. Alexander, *X-Ray Diffraction Procedures for Polycrystalline and Amorphous Materials*, New York, J. Wiley, ed. III, 1962.
423. W. H. Beathie and C. Booth, *J. Appl. Polym. Sci.* **7,** 507 (1963).
424. H. L. Hsieh, *J. Polym. Sci.* **A3,** 191 (1965).
425. I. Ya. Poddubnyi and E. G. Ehrenburg, *Vysokomol. Soed.* **4,** 961 (1962).
426. J. C. W. Chien, *J. Polym. Sci.* **A1,** 425 (1963).
427. J. C. W. Chien, *J. Polym. Sci.* **A1,** 1839 (1963).
428. E. Ceauşescu, P. Lebădă, E. Mihăilescu and V. Guzic, *Mat. Plast.* **3,** 127 (1966).
429. K. B. Piotrovskii, *Zh. Vses. Khim. Otd. im Mendeleeva*, **11,** 3 (1966).
430. G. Scott, *Chem. Ind.* **2,** 272 (1966).
431. T. Kemperman, *Rev. Gen. Caout.* **40,** 406 (1963).
432. J. C. Ambelang and J. R. Shelton, *Rubber Chem. Technology*, **36,** 1497 (1963).
433. C. E. Brockway, Fr. Pat., 1172383 (1956).
434. Brit. Pat. 872283 (1961).
435. Ital. Pat., 605067 (1959).
436. H. Groene and G. Pampus, Ger. Pat., 1120695 (1961).

Part II

1. E. Ceauşescu, S. Bittman, B. Hlevca, I. Florescu, E. Mihăilescu and I. Ciută, *Rev. Chim.* **12,** 284 (1961).
2. E. Ceauşescu, S. Bittman, I. Florescu and L. Buşilă, Paper presented at the ICECHIM Symposium, Bucharest, 1963.
3. E. Ceausescu, S. Bittman, I. Florescu, V. Fieroiu and C. Ivaşcu, Paper presented at the Centennial Symposium of the Academy of the Romanian Socialist Republic, Bucharest, 1966.
4. E. Ceauşescu, S. Bittman, I. Florescu and P. Lebădă, Paper presented at the 36th Congress of Industrial Chemistry, Brussels, 1966.
5. H. de Vries, *Rec. Trav. Chim.* **80,** 866 (1961).
6. G. Natta, P. Corradini and G. Allegra, *J. Polym. Sci.* **51,** 399 (1961).
7. P. Cossee, *Trans. Faraday Soc.* **58,** 1226 (1962).
8. E. Schoenberg, D. L. Chalfort and R. H. Meyer, *Rubb. Chem. Tech.* **37,** 103 (1964).
9. E. Schoenberg, D. L. Chalfort and T. L. Hanlon, *Adv. Chem. Ser.* **52,** 1215 (1966).
10. E. Ceauşescu, S. Bittman, I. Florescu and R. Bordeianu, *Mat. Plast.* **2,** 315 (1965).
11. B. Gerhard *et al.*, Ger. Pat., 1301487 (1969).
12. O. Hamrik *et al.*, Czech. Pat., 115254 (1967).
13. W. M. Saltman and B. Wargortz, Fr. Pat., 1230794 (1960).

14. E. CEAUŞESCU *et al.*, Rom. Pat., 71113 (1979).
15. G. F. COATES and M. L. H. GREEN, *Organometallic Compounds*, Methuen, London, 1963, chapter 6.
16. I. URSU, *La résonance paramagnetique electronique*, Dunod, Paris, 1967.
17. E. A. SHILOV, A. K. ZEFIROVA and N. N. TIKHOMIROVA, *Zh. Fiz. Khim.* **33,** 2113 (1959).
18. A. K. ZEFIROVA, N. N. TIKHOMIROVA and A. E. SHILOV, *Dokl. Akad. Nauk SSSR*, **132,** 1082 (1960).
19. P. E. ALLEN, J. K. BROWN and R. M. OBAID, *Trans. Faraday Soc.* **59,** 1808 (1963).
20. H. MAKI and E. V. RANDALL, *J. Am. Chem. Soc.* **82,** 4019 (1960).
21. E. H. ADEMA, H. J. M. BARTELINK and J. SMIDT, *Rec. Trav. Chim.* **80,** 173 (1961).
22. E. H. ADEMA, H. J. M. BARTELINK and J. SMIDT, *Rec. Trav. Chim.* **81,** 73 (1962).
23. H. J. M. BARTELINK, H. BOS, J. SMIDT, C. M. VRINSEN and E. H. ADEMA, *Rec. Trav. Chim.* **81,** 225 (1962).
24. A. E. SHILOV, A. K. SHILOVA and V. N. BOBKOV, *Vysokomol. Soed.* **4,** 11 (1962).
25. A. TKAC, *Coll. Czech. Chem. Commun.* **33,** 1629 (1968).
26. E. ANGELESCU, C. NICOLAU and Z. SIMON, *J. Am. Chem. Soc.* **88,** 3910, (1966).
27. T. S. DJABIEV, R. D. SABIROVA and A. E. SHILOV, *Kinetika i Kataliz*, **5,** 441 (1964).
28. K. HIRAKI, S. KANOKA and H. HIRAI, *J. Polym. Sci., Polym. Lett.* **10,** 199 (1972).
29. H. HIRAI, K. HIRAKI, I. NOGUCHI, T. INOUE and S. MAKISHINA, *J. Polym. Sci.* **A1,** 8, 2392 (1970).
30. H. HIRAI, K. HIRAKI, I. NOGUCHI and S. MAKISHINA, *J. Polym. Sci.* **A1,** 8, 147 (1970).
31. M. TAKEDA, K. IIMURA, Y. NOZAWA, M. HISATOME and N. KOIDE, *J. Polym. Sci.* **C23,** 741 (1968).
32. E. E. BENGEL', N. G. MAKSIMOV, V. M. MASTIKHIN, K. G. ARTAMANOVA, V. F. AMEFENKO and V. A. ZAKHAROV, *Kinetika i Kataliz*, **16,** 1015 (1975).
33. N. G. MAKSIMOV, E. G. KUSHNAREVA, V. A. ZAKHAROV, V. F. AMEFENKO, P. A. ZHDAN and YU. I. ERMAKOV, *Kinetika i Kataliz*, **15,** 738 (1974).
34. Z. P. ARKHIPOVA, V. K. BADAEV and B. V. EROFEYEV, *Dokl. Akad. Nauk SSSR*, **183,** 1317 (1968).
35. Goodyear Tire & Rubber, Brit. Pat., 1017889 (1966).
36. J. BOOR, Jr., *J. Polym. Sci.* **C1,** 257 (1963); **B3,** 7 (1965); **A3,** 995 (1965).
37. A. D. CAUNT, *J. Polym. Sci.* **C4,** 49 (1963).
38. H. W. COOVER, Jr. and F. B. JOYNER, *J. Polym. Sci.* **A3,** 2407 (1965).
39. R. L. MCCONELL, M. A. MCCALL, J. O. CASH, Jr., F. B. JOYNER and H. W. COOVER, Jr., *J. Polym. Sci.* **A3,** 2135 (1965).
40. J. AMBROŽ and O. HAMRIK, *Coll. Czech. Chem. Commun.* **28,** 2550 (1963).
41. H. KOHN and S. HORNE, U.S. Pat., 3165503 (1965).
42. BAYER, A. G., Fr. Pat., 2000038 (1969).
43. M. GIPPIN, *J. Appl. Polym. Sci.* **14,** 1807 (1970).
44. Michelin & Co., Brit. Pat., 1182935 (1969).
45. Goodyear Tire & Rubber Co., Brit. Pat., 1043439 (1966).
46. R. S. KIRPICHEVA, F. D. RADENKOV, L. I. PETROV and M. KH. MIKHAILOV, *Plast. Massy*, **7,** 3 (1971).
47. K. MORRI *et al.*, Ger. Pat., 1800713 (1969).
48. Goodyear Tire & Rubber Co., Brit. Pat., 1004665 (1965).
49. Firestone Tire & Rubber Co. U.S. Pat., 3523114 (1970).
50. J. D. D'IANNI, *Kautsch. Gummi Kunst.* **19,** 1276 (1966).
51. P. COSSEE, *J. Catal.* **3,** 80 (1964).
52. K. VESELÝ, J. AMBROŽ, R. VILIM and O. HAMRIK, *J. Polym. Sci.* **55,** 25 (1961).
53. G. A. RAZUVAEV, K. S. MINSKER and R. P. CHERNOVSKAYA, *Dokl. Akad. Nauk SSSR*, **147,** 636 (1962).
54. R. P. CHERNOVSKAYA, K. S. MINSKER and G. A. RAZUVAEV, *Vysokomol. Soed.* **6,** 1656 (1964).
55. K. S. MINSKER and V. K. BYKOVSKII, *Vysokomol. Soed.* **2,** 535 (1960).
56. E. CEAUŞESCU, S. BITTMAN and E. BUZDUGAN, Paper presented at the 5th National Conference on Physical Chemistry, Bucharest, Sept. 1976.
57. W. COOPER, D. E. EAVES, G. D. T. OWEN and G. VAUGHAM, *J. Polym. Sci.* **C4,** 211 (1963).
58. W. COOPER, R. K. SMITH and A. STOKES, *J. Polym. Sci.* **B4,** 309 (1966).

59. E. Bonitz, *Chem. Ber.* **88,** 742 (1955).
60. R. D. Bushick and R. S. Stearns, *J. Polym. Sci.* **A1,** 4, 215 (1966).
61. I. V. Nicolescu and E. Angelescu, *J. Polym. Sci.* **A1,** 3, 1227 (1965).
62. I. V. Nicolescu and E. Angelescu, *J. Polym. Sci.* **A1,** 4, 2963 (1966).
63. A. H. Adema, H. J. M. Bartelink and J. Smidt, *Rec. Trav. Chim.* **80,** 173 (1961).
64. M. B. Smith, *J. Organomet. Chem.* **22,** 273 (1970).
65. M. B. Smith, *J. Organomet. Chem.* **70,** 13 (1974).
66. H. E. Swift, C. P. Poole, Jr. and J. F. Itzel, *J. Phys. Chem.* **68,** 2509 (1964).
67. R. M. Hamilton, R. McBeth and W. Bekbrede, *J. Am. Chem. Soc.* **75,** 2881 (1953).
68. E. Ceauşescu *et al.*, Rom. Pat., 53585 (1970).
69. E. Ceauşescu *et al.*, Rom. Pat., 56864 (1971).
70. E. Ceauşescu *et al.*, Rom. Pat., 69606 (1978).
71. E. Ceauşescu, S. Bittman, I. Florescu, V. Fieroiu, M. Labă, N. Iacob, A. Cornilescu, M. Popescu, E. Nicolescu, E. G. Badea, E. Buzdugan and V. Gruber, Paper presented at the National Congress of Chemistry, Bucharest, 1978.
72. E. Ceauşescu, S. Bittman, V. Fieroiu, I. Florescu and C. Ivaşcu, *Rev. Roumaine Chim.* **10,** 3 (1965).
73. E. Ceauşescu *et al.*, Rom. Pat., 43694 (1964).
74. A. A. Korotkov, E. N. Marandjeva and Z. A. Khrenova, *Isoprene Polymerisation with Complex Catalysts* (Russ.), A. A. Korotkov, ed., Khimiya, Leningrad, 1964, p. 41.
75. Yu. B. Monakov, N. N. Minchenkova and R. S. Rafikov, *Dokl. Akad. Nauk SSSR*, **236,** 1151 (1977).
76. N. N. Minchenkova, *Research in the Field of Macromolecular Chemistry*, Bashk. branch of the USSR Acad. Sci., Ufa, 1977.
77. E. Ceauşescu, C. Ivaşcu, L. Petcu, M. Mărgineanu and T. Bălan, Paper presented at the Macromolecular Chemistry Symposium, Bucharest, 1965.
78. E. Ceauşescu, C. Ivaşcu, M. Mărgineanu, T. Bălan and C. Popescu, Paper presented at the 1st National Conference on Macromolecular Compounds, Iassy, 1968.
79. A. A. Korotkov, V. A. Kormer, M. A. Krupyshev and D. P. Ferringer, *Isoprene Polymerisation with Complex Catalysts* (Russ.), A. A. Korotkov, ed., Khimiya, Leningrad, 1964, p. 3.
80. V. A. Grechanovskii and I. Ya. Poddubnyi, *Vysokomol. Soed.* **17A,** 2154 (1975).
81. V. A. Grechanovskii, G. A. Martinovskii and I. M. Sapozhnikova, *Vysokomol. Soed.* **17A,** 514 (1975).
82. V. A. Grechanovskii and I. Ya. Poddubnyi, *Vysokomol. Soed.* **12B,** 875 (1974).
83. V. A. Grechanovskii, L. S. Ivanova and I. Ya. Poddubnyi, *Vysokomol. Soed.* **15A,** 889 (1973).
84. V. A. Grechanovskii and I. Ya. Poddubnyi, *Dokl. Akad. Nauk SSSR*, **197,** 643 (1971).
85. A. A. Korotkov, N. A. Chesnokova and M. A. Krupyshev, *Isoprene Polymerisation with Complex Catalysts* (Russ.), A. A. Korotkov, ed., Khimiya, Leningrad, 1964.
86. A. A. Korotkov, A. A. Vasil'ev, V. V. Prokof'ev and N. P. Timofeeva, *Isoprene Polymerisation with Complex Catalysts* (Russ.), A. A. Korotkov ed., Khimiya, Leningrad, 1964.
87. P. A. Kirpichnikov, L. A. Averko-Antonovich and Yu. O. Averko-Antonovich, *Chemistry and Technology of Synthetic Rubber* (Russ.), Khimiya, Leningrad, 1975.
88. E. Ceauşescu, S. Bittman, I. Florescu, V. Fieroiu, P. Lebădă, N. Tudorache and R. Bordeianu, Paper presented at the 12th Latin-American Congress of Chemistry, Quito, Ecuador, Sept. 1976.
89. V. A. Kormer and V. A. Vasil'ev, *Synthetic Rubber* (Russ.), I. V. Garmonov ed., Khimiya, Leningrad, 1976, p. 207.
90. M. V. Bulgakova and Ya. M. Rozinoer, Monomers and macromolecular compounds (Russ.), *Tr. Voronezh. Gosud. Univ.* **73,** 114 (1969).
91. Yu. A. Litvin, V. M. Nepyshnevskii and G. M. Sinaiskii, *Kauch. Rezina*, (2), 1 (1962).
92. I. M. Sapozhnikov, V. A. Zhukov, N. G. Pavlov and A. S. Estrin, *Kauch. Rezina*, (2), 5 (1976).
93. B. O. Reikhsfel'd and L. N. Erkova, *Equipments for Organic Syntheses and Synthetic Rubber Manufacture* (Russ.), Khimiya, Leningrad, 1974.

94. E. CEAUŞESCU, P. LEBĂDĂ, V. APOSTOL and R. MUNTEANU, *Mat. Plast.* **9,** 313 (1972).
95. E. N. MARANDJEVA, A. V. ZAK and A. S. ESTRIN, *Prom. Sint. Kauch.* (8), 7 (1969).
96. I. V. GARMONOV, V. A. KORMER and ZH. VSESOYUZ. *Khim. Obsch. im. Mendeleeva,* **19,** 628 (1974).
97. A. A. KOROTKOV, N. N. CHESNOKOVA and M. A. KRUPYSHEV, *Isoprene Polymerisation with Complex Catalysts* (Russ.), A. A. KOROTKOV, ed., Khimiya, Leningrad, 1964, p. 14.
98. V. M. BREITMAN and A. V. ZAK, *Zh. Prikl. Khim.* **44,** 786 (1971).
99. A. V. ZAK, V. A. LAVROV, ZH. B. PEKEL'NAYA, M. S. PERFIL'EVA, M. D. ROGOZKINA and P. P. SHPAKOV, *Zh. Prikl. Khim.* **50,** 142 (1977).
100. A. V. ZAK, B. A. PERLIN, V. M. BREITMAN, V. E. GURARI, V. A. LAVROV and S. L. PODVAL'NYI, *Zh. Prikl. Khim.* **48,** 1878 (1975).
101. B. F. REITSES, V. O. REIKHSFEL'D, A. V. ZAK and I. M. ABRAMSON, *Prom. Sint. Kauch.* (9), (3) (1979).
102. B. F. REITSES, A. V. ZAK, I. M. ABRAMSON and V. O. REIKHSFEL'D, *Zh. Prikl. Khim.* **51,** 2549 (1978).
103. B. F. REITSES, I. M. ABRAMSON, A. V. ZAK, S. I. LUZHABSKII, B. A. PERLIN and YU. V. POPOV, *Prom. Sint. Kauch.* (12), 3 (1974).
104. A. V. ZAK and E. G. ERUSSALIMSKII, *Prom. Sint. Kauch.* (5), 13 (1974).
105. E. CEAUŞESCU, S. BITTMAN, V. FIEROIU, P. LEBĂDĂ, M. LABĂ, V. ABABI, C. DRĂGUS and E. MIHĂILESCU, Paper presented at the International Symposium on Polyisoprene, Moscow, 1972.
106. V. A. GRECHANOVSKII and I. YA. PODDUBNYI, *Synthetic Rubber* (Russ.), I. V. GARMONOV, ed., Khimiya, Leningrad, 1976, p. 72.
107. I. YA. PODDUBNYI and V. A. GRECHANOVSKII, *Kauch. Rezina,* (7), 10, (1972).
108. I. YA. PODDUBNYI and ZH. VSESOYUZ. *Khim. Obsch. im. Mendeleeva,* **19,** 638 (1974).
109. E. CEAUŞESCU, S. BITTMAN, A. CORNILESCU, E. NICOLESCU, M. POPESCU, E. G. BADEA, E. BUZDUGAN and V. GRUBER, Paper presented at the ICECHIM Symposium, Bucharest, 1978.
110. I. YA. PODDUBNYI and E. G. ERENBURG, *Vysokomol. Soed.* **4,** 961 (1962).
111. R. E. HARRINGTON and B. H. ZIMM, *J. Polym. Sci.* **A2,** 6, 294 (1968).
112. A. G. LIAKUMOVICH and N. N. SMIRNOV, *Prom. Sint. Kauch.* (7), 4 (1969).
113. N. G. TURASH, L. M. POSPELOVA and E. M. SIRE, *Prom. Sint. Kauch.* (3), 11(1979).
114. V. A. GRECHANOVSKII, I. P. DMITRIEVA, E. P. PISKAREVA and I. YA. PODDUBNYI, *Kauch. Rezina,* (3), 8 (1977).
115. V. K. NEL'SON and G. S. SOLODOVNIKOVA, *Kauch. Rezina,* (4), 4 (1972).
116. N. F. KOVALEV, G. A. TIHOMIROVA and A. S. ESTRIN, *Kauch. Rezina,* (7) 6 (1975).
117. S. I. VOL'FSON, M. G. KARP, F. A. GARRIFULIN and A. G. LIAKUMOVICH, *Kauch. Rezina,* (11), 9 (1978).
118. V. A. GRECHANOVSKII, *Kauch. Rezina,* (1), 8 (1976).
119. V. A. GRECHANOVSKII, E. M. SIRE, L. M. POSPELOVA and S. A. DUL'KINA, *Kauch. Rezina,* (4), 6 (1975).
120. V. A. GRECHANOVSKII, L. S. IVANOVA, I. YA. PODDUBNYI and E. M. SIRE, *Kauch. Rezina,* (12), 18 (1972).
121. V. A. GRECHANOVSKII, L. S. IVANOVA and I. YA. PODDUBNYI, *Kauch. Rezina,* (6), 5 (1972).
122. E. M. SIRE, L. M. POSPELOVA and T. N. FIMUTINA, *Kauch. Rezina,* (8), 6 (1972).
123. V. A. GRECHANOVSKII, L. S. IVANOVA and N. F. KOVALEV, *Kauch. Rezina,* (2), 3 (1975).
124. G. S. SAFRONOVA, V. G. ZHAKOVA, N. F. KOVALEV and B. K. KARMIN, *Prom. Sint. Kauch.* (12), 6, (1976).
125. I. V. GARMONOV, A. V. ZAK, V. A. LAVROV, A. K. LIL'EVA, B. A. PERLIN and P. P. SHPAKOV, *Zh. Prikl. Khim.* **49,** 2661 (1976).
126. I. V. GARMONOV, A. K. LIL'EVA, A. V. ZAK and B. A. PERLIN, *Kauch. Rezina,* (2), 6 (1977).
127. M. P. KULIK, E. P. PISTUN and T. S. PODOL'SKII, *Prom. Sint. Kauch.* (2), 9 (1977).
128. A. S. LIVSHITSYN, N. G. PAVLOV and I. M. SAPOZHNIKOV, *Prom. Sint. Kauch.* (2), 11 (1978).
129. A. P. BADYEV *et al.,* U.S.S.R. Pat., 530034 (1976).
130. V. SVOBODA, O. VILIM and A. KUBICEC, *Coll. Czech. Chem. Commun.* **28,** 2310 (1963).
131. G. P. BELONOVSKAYA, B. A. DOLGOPLOSK and ZH. D. CHESNOVA, *Vysokomol. Soed.* **4,** 161 (1962).

132. Lee Lieng-Huang, C. L. Hanci and L. G. Engel, *J. Appl. Polym. Sci.* **10,** 1699 (1966).
133. W. C. Rao, H. Winn and J. R. Shelton, *Ind. Eng. Chem.* **44,** 574 (1952).
134. N. Uri, *Antixoidation and Antioxidants*, vol. 1, N. O. Lundberg, ed., Interscience, New York, 1961.
135. S. E. Horne, U.S. Pat., 2962488 (1960).
136. L. Rosik and V. Svoboda, *Coll. Czech. Chem. Commun.* **31,** 1513 (1966).
137. E. Ceauşescu, V. Fieroiu, P. Lebădă, G. Tripcovici, E. Simionescu and V. Guzic, *Mat. Plast.* (2), 320 (1965).
138. Goodrich Gulf, Brit. Pat., 1172383 (1969).
139. J. R. Shelton and W. L. Cox, *Ind. Eng. Chem.* **45,** 392, 397 (1953).
140. J. K. Aiken, *New Scientist*, **15,** 397 (1962).
141. J. R. Shelton, *Polymer Stabilization*, W. L. Hawkins, ed., Wiley, New York, 1972, p. 29.
142. A. S. Kuzminskii, V. D. Zaitsev and N. N. Lezhner, *Vysokomol. Soed.* **4,** 1682 (1962).
143. J. C. Petersen, *Ind. Eng. Chem.* **41,** 924 (1949).
144. M. Haruta, K. Kageno and S. Harada, *Makromol. Chem.* **167,** 95 (1973).
145. J. R. Shelton, *Rubb. Chem. Technol.* **45,** 359 (1972).
146. H. U. Voight, A. Stefkova and W. Scheele, *Kautsch. Gummi Kunst.* **23,** 359 (1970).
147. J. C. Amberlang, R. H. Kline, O. M. Loentz, C. R. Parks, C. N. Adelin and J. R. Shelton, *Rubb. Chem. Technol.* **36,** 1497 (1963).
148. N. Uri, *Antioxidation and Antioxidants*, vol. 1, W. O. Lundberg, ed., Interscience, New York, 1961.
149. L. Rosik, *J. Polym. Sci., Polym. Lett.* **5,** 1083 (1967).
150. K. B. Piotrovskii, *Vysokomol. Soed.* **14B,** 302 (1972).
151. W. E. McCormick, *Rubb. Chem. Technol.* **45,** 627 (1972).
152. V. P. Kirpichev, L. P. Andreeva and A. I. Yakubchik, *Zh. Prikl. Khim.* **47,** 2322 (1974).
153. R. E. Morris *et al.*, U.S. Pat., 3177165 (1965).
154. H. U. Voight and F. U. Krüger, *Kautsch. Gummi Kunst.* **24,** 399 (1971).
155. H. U. Voight and F. U. Krüger, *Kautsch. Gummi Kunst.* **25,** 133 (1972).
156. E. Ceauşescu, V. Fieroiu, B. Calmanovici and E. Simionescu, *Mat. Plast.* **8,** 2 (1971).
157. E. Ceauşescu, V. Fieroiu, B. Calmanovici, E. Simionescu and L. Fenici, *Mat. Plast.* **6,** 255 (1969).
158. G. V. Vinogradov and L. I. Ivanova, *Kauch. Rezina*, (1), 15 (1968).
159. W. Hofmann, *Gummi Asbest. Kunst.* **20,** 602 (1967).
160. B. N. Leyland, *Prog. Rubb. Technol.* **36,** 19 (1972).
161. W. E. McCormick, *Rubb. Chem. Technol.* **44,** 512 (1971).
162. J. D. D'Ianni, *Rev. Gen. Caout. Plast.* **47,** 844 (1970).
163. J. R. Dunn, *Rubb. Chem. Technol.* **47,** 960 (1974).
164. D. Ryshavy, L. Balaban, V. Slavic and I. Ruja, *Vysokomol. Soed.* **3,** 1110 (1961).
165. D. Ryshavy, *Polymer*, **8,** 449 (1967).
166. D. Reichenbach and W. Scheele, *Kautsch. Gummi Kunst.* **23,** 141 (1970).
167. E. Ceauşescu, S. Bittman and E. G. Badea, Paper presented at the 5th National Conference on Physical Chemistry, Bucharest, Sept. 1977.
168. M. B. Hughin and G. J. Knight, *Makromol. Chem.* **152,** 67 (1972).
169. T. V. Fedorova and S. M. Kavun, *Kauch. Rezina*, (9), 13 (1974).
170. G. N. Petrov and K. B. Piotrovskii, *Vysokomol. Soed.* **17B,** 429 (1975).
171. V. P. Kirpichev and K. B. Piotrovskii, *Kauch. Rezina*, (7), 20 (1972).
172. G. I. Bebikh and V. P. Saraeva, *Vysokomol. Soed.* **18,** 461 (1976).
173. V. P. Kirpichev and A. I. Yakubchik, *Vysokomol. Soed.* **11A,** 2293 (1969).
174. V. P. Kirpichev and A. I. Yakubchik, *Vysokomol. Soed.* **10A,** 2347 (1968).
175. J. T. Gregory *et al.*, U.S. Pat., 3177166 (1965).
176. B. N. Leyland, *IRI J.* **5,** 51 (1971).
177. S. M. Hirshfield and E. R. Allen, *J. Polym. Sci.* **39,** 544 (1959).
178. *Chem. Eng. News*, **37** (3), 50 (1959).
179. C. T. Winchester, *Ind. Eng. Chem.* **51,** 19 (1959).
180. *Rubb. Plast. Age*, **41,** 1168 (1960).
181. C. L. Beal and L. Basel, *Chem. Eng. Progr.* **57,** (5), 50 (1961).
182. J. W. Dunning and S. Baer, *Chem. Eng. Progr.* **57** (5), 53 (1961).

183. K. Hattori and Y. Komeda, U.S. Pat., 3583967 (1971).
184. H. Schnoriry and G. Pampus, Brit. Pat., 1232051 (1971).
185. O. W. Burke, U.S. Pat., 3622127 (1972).
186. E. Ceauşescu *et al.*, Rom. Pat., 42812 (1962).
187. E. Ceauşescu *et al.*, Rom. Pat., 49888 (1965).
188. N. Gaylord and H. Mark, *Encyclopedia of Polymer Science and Technology*, vol. 14, Wiley, New York, 1971, p. 149.
189. Hung Yu Chen, *Anal. Chem.* **34,** 1134 (1962).
190. Hung Yu Chen, *Anal. Chem.* **34,** 1973 (1962).
191. F. C. Stehling and K. W. Bartz, *Anal. Chem.* **38,** 467 (1966).
192. M. A. Golub, S. A. Fuqua and N. S. Bhacca, *J. Am. Chem. Soc.* **84,** 4981 (1962).
193. D. J. Worsfold and S. Bywater, *Can. J. Chem.* **42,** 2884 (1964).
194. F. Schué, *Bull. Soc. Chem. France*, **14,** 980 (1965).
195. Hung Yu Chen, *J. Polym. Sci.* **B4,** 891 (1966).
196. A. L. Buchachenko, A. L. Kovarakii and A. M. Vasserman, *Adv. Polym. Sci.*, Z. A. Rogovin, ed., Wiley, New York, 1974, p. 37.
197. G. P. Rabold, *J. Polym. Sci.* **A1,** 7, 1203 (1969).
198. P. L. Kumler and R. E. Boyer, *Macromol.* **9,** 93 (1976).
199. R. E. Boyer, *Rubb. Chem. Technol.* **36,** 1303 (1963).

Index